Hydrocarbon Pollution and its Effect on the Environment

*Edited by Muharrem Ince
and Olcay Kaplan Ince*

Published in London, United Kingdom

IntechOpen

Supporting open minds since 2005

Hydrocarbon Pollution and its Effect on the Environment
http://dx.doi.org/10.5772/intechopen.81556
Edited by Muharrem Ince and Olcay Kaplan Ince

Contributors
Hugo Saldarriaga-Noreña, Mario Alfonso Murillo-Tovar, Josefina Vergara-Sánchez, Rebecca López-Márquez, Jorge Antonio Guerrero-Alvarez, Mónica Ivonne Arias -Montoya, Hareef Baba Shaeb K, Manish Srivastava, Anamika Srivastava, Varun Rawat, Anjali Yadav, Peramaiyan Rajendran, Abdullah. M. Alzahrani, Muharrem Ince

Notice
Statements and opinions expressed in the chapters are these of the individual contributors and not necessarily those of the editors or publisher. No responsibility is accepted for the accuracy of information contained in the published chapters. The publisher assumes no responsibility for any damage or injury to persons or property arising out of the use of any materials, instructions, methods or ideas contained in the book.

First published in London, United Kingdom, 2019 by IntechOpen
IntechOpen is the global imprint of INTECHOPEN LIMITED, registered in England and Wales, registration number: 11086078, 7th floor, 10 Lower Thames Street, London,
EC3R 6AF, United Kingdom
Printed in Croatia

British Library Cataloguing-in-Publication Data
A catalogue record for this book is available from the British Library

Additional hard and PDF copies can be obtained from orders@intechopen.com

Hydrocarbon Pollution and its Effect on the Environment
Edited by Muharrem Ince and Olcay Kaplan Ince
p. cm.
Print ISBN 978-1-78984-420-7
Online ISBN 978-1-78984-421-4
eBook (PDF) ISBN 978-1-83968-031-1

We are IntechOpen,
the world's leading publisher of
Open Access books
Built by scientists, for scientists

4,400+
Open access books available

118,000+
International authors and editors

130M+
Downloads

Our authors are among the

151
Countries delivered to

Top 1%
most cited scientists

12.2%
Contributors from top 500 universities

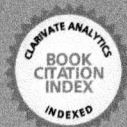

CLARIVATE ANALYTICS
BOOK
CITATION
INDEX
INDEXED

WEB OF SCIENCE™

Selection of our books indexed in the Book Citation Index
in Web of Science™ Core Collection (BKCI)

Interested in publishing with us?
Contact book.department@intechopen.com

Numbers displayed above are based on latest data collected.
For more information visit www.intechopen.com

Meet the editors

Dr. Muharrem Ince received his Ph.D. degree in analytical chemistry at Firat University, Turkey in 2008. He worked at Mus Alparslan University, Turkey from 2009 to 2012 as an assistant professor. He has been working at Munzur University since 2012. From 2013 to 2016, he served as Head of Department of Chemical Engineering at the Munzur University. He has been an "Editorial Board Member" of several international journals. Currently, he is an associate professor at the Munzur University Rare Earth Elements Research and Application Center. He has authored and co-authored more than 26 papers that have been published in respectable international journals. He is an expert in analytical method development, spectroscopic and chromatographic techniques, food analysis and toxicology, nanoscience, and nanotechnology.

Dr. Olcay Kaplan Ince received her BSc degree from Hacettepe University and her PhD degree in analytical chemistry from Firat University in Turkey in 2008. She became a research analytical chemist at the Munzur University in 2009 in the Food Engineering Department. She was Head of the Department of Food Engineering at the Munzur University from 2014 to 2015. She currently serves as an Associate Editor of the International Journal of Pure and Applied Sciences since 2015. She is the author of more than 30 papers published in journals with good impact factors in their area and her research areas include trace and toxic element analysis, analytical chemistry, instrumental analysis, problem solving in analytical chemistry, food science and chromatography, and nanoscience and citotoxicology.

Contents

Preface

This book is titled "Hydrocarbon Pollution and its Effect on the Environment". The book provides an overview on hydrocarbon identification, contamination, and measurement techniques. "Hydrocarbon Pollution and its Effect on the Environment" takes a broad view of the subject and integrates a wide variety of approaches. This book attempts to address the needs of graduate and postgraduate students, as well as chemists and other professionals or readers interested in food, soil, water, and air pollution. The aim of this book is to explain important studies, and compare and develop the new and groundbreaking measurement techniques. Written by leading experts in their respective areas, the book is recommended to professionals interested in environmental and human health because it provides specific and comprehensive examples.

Dr. Muharrem Ince and Dr. Olcay Kaplan Ince
Munzur University Rare Earth Elements Research and Application Center,
Turkey

Introductory Chapter: Sources, Health Impact, and Environment Effect of Hydrocarbons

Muharrem Ince and Olcay Kaplan Ince

1. Introduction

Pollution control and environmental protection have become a worldwide issue of concern. The aliphatic hydrocarbons (AHs), aromatic hydrocarbons (ArHs) such as benzene and toluene, and polycyclic aromatic hydrocarbons (PAHs), including benzo[a]anthracene, benzo[ghi]pyrilene, and benzo[a]pyrene, are persistent organic pollutants (POPs) for ecosystem. These hazardous pollutants are risky because of mutagenic, carcinogenic, immunotoxic, and teratogenic effects. These components threaten all life forms ranging from microorganisms to humans when they are released into the environment especially via human activities. The aim of this study is to provide up-to-date information on the various hydrocarbons present in the environment, routes of exposure, and their adverse impact on environment and human health. There are two major categories that contain hydrocarbons; these are aliphatic and aromatic compounds (**Figure 1**). While aromatic hydrocarbons contain at least one benzene ring, the other group called as nonaromatic or aliphatic does not contain it. The basic structure that forms aromatic hydrocarbons is the benzene ring. On the other hand, petroleum hydrocarbons (PHCs) comprised of carbon and hydrogen atoms which are organic compounds. They have varying structural configurations with physical and chemical characteristics. They can be broadly classified as gasoline range organics (GROs) and diesel range organics (DROs). The first group that is called GROs comprises monoaromatic hydrocarbons including toluene, benzene, and ethylbenzene. This category has short-chain alkanes ranging from 6 to 10 C. The second group that is called DROs has longer C-chain alkanes from 10 to 40 C, and this category contains hydrophobic chemicals including polycyclic aromatic hydrocarbons (PAHs) [1, 2]. These compounds, in contaminated ecosystem, are considered to be one of the most stable hydrocarbon forms. The PAH molecular weight is the main factor to determine their origin's level in earth. There are two PAH sources: natural and anthropogenic. Both sources are important and remarkable. Because of natural and anthropogenic activities, these pollutants are irregularly distributed throughout various levels and locations to all over the world. Various studies have revealed that PAHs have carcinogenic, teratogenic, and mutagenic effect on human health [3, 4]. The main skeleton of these compounds, classified as organic pollutants, consists of two or more benzene rings. The extensive nonpolar contaminants are detected in petrochemical products including coal, oil, and tar. Another significant source of hydrocarbons is also incomplete combustion [5–8]. According to researchers, because of ecotoxicological risks and potential sources, 26 AHs and 16 PAHs causing concerns for ecosystem are categorized as carcinogen or mutagen by the United States Environmental Protection Agency (USEPA). These ecotoxicological compounds include benzo[a]pyrene, benz[a]anthracene, etc.

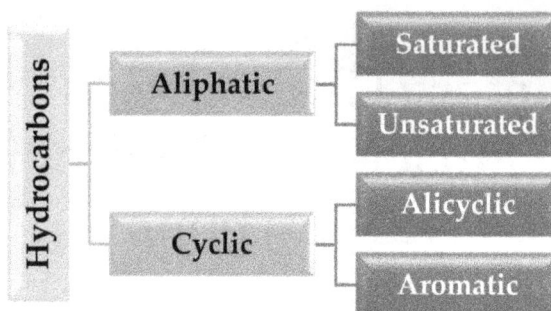

Figure 1.
Hydrocarbon classification [42].

The USEPA mentioned that there are 126 major pollutants in the environment, 25 of them were threatening and 5 were extremely dangerous for the environment. For example, some health institutions including the USEPA and International Agency for Research on Cancer (IARC) mentioned that benzo[*a*]pyrene that is a member of PAHs is carcinogenic for animals and humans [7, 9–11].

2. Sources of hydrocarbons

The major hydrocarbon sources are petroleum and petroleum combustion; however, their emission sources can be classified as phytogenic (natural), petrogenic, and pyrogenic. To recognize pollutant type and migration, circumstances play a key role for their origin [12]. Hydrocarbons can enter to the environment via dispersion, evaporation, dissolution, adsorption, and other processes including petroleum and petroleum combustion [13, 14]. Petrogenic sources generally pollute groundwater and threaten the environment because petrogenic source products including lubricants and fuels leak from the tanks and release into the environment [15]. The USEPA specified 16 priority PAHs in a petroleum source, namely, alkylated naphthalene, dibenzothiophene, fluorene, phenanthrene, and chrysene series [16]. The pyrogenic PAHs are produced during the fuel combustion because there are suitable conditions that are high temperature and absence of oxygen. Also, pyrolysis of fat and incomplete combustion besides power plants are the most prominent hydrocarbon sources [17]. Hydrocarbons and their derivatives are a significant environmental concern due to their extensive use and toxic mechanism action, and these products are highly available in aquatic medium [18, 19]. Industrial activities and chemical plants produce PAHs, and they are considered as petrogenic and natural PAH sources [20]. During fat pyrolysis and incomplete combustion processes, anthropogenic emissions of PAHs are released into the environment [7, 8]. On the other hand, PAH sources were classified as natural, industrial, domestic, agricultural, and mobile by Ravindra et al. [21]. Hydrocarbons are usually generated by various sources including wildfires, oil seepages, volcanic activities, and other sources. Moreover, these natural hydrocarbons are mainly produced during organic material chemical conversions in microorganisms, fungi, plants, sediments, etc. [16, 22–24].

3. Health threat and environmental impact assessment

Recent studies have recognized the effects of toxicity, mutagenicity, and carcinogenicity of hydrocarbons. Increasing contamination level of these pollutants

```
┌─────────────────────────────────────────────────┐
│  Environmental impact assessment stages          │
└─────────────────────────────────────────────────┘
        │
        │────── Hazard identification
        │
        │────── Measuring of the probability of the consequences
        │
        │────── Measuring of the probability of the consequences
        │
        │────── Measuring of the magnitude of the consequences
        │
        └────── Evaluating the significance of a risks
```

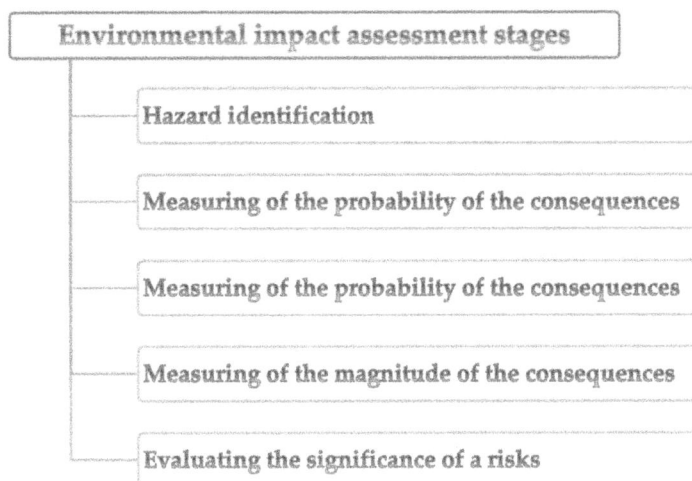

Figure 2.
Environmental impact assessment stages.

in environment especially in aquatic media is a significant environmental concern because they are used frequently and show environmental toxic effects [25–28]. The USEPA and World Health Organization (WHO) classified PAHs and total petroleum hydrocarbons (TPHs) as POP groups in marine and coastal environment [29, 30]. The most of PAHs have been banned by health authorities due to their long half-life, wide distribution, and high bioaccumulation in the food chain, as well as their potential for toxicity to humans, because these compounds are highly lipid soluble and these toxic chemicals can bioaccumulate from environment to the gastro-intestinal tract of mammals [25, 31]. When animals and humans are exposed to hydrocarbons, it is probable that they have various health problems because they are vulnerable and endangered against these components. Research on some hydrocarbons including benzo[a]pyrene, pyrene, and benzo[a]anthracene have revealed that these compounds have carcinogenic and mutagenic effect [7, 8, 11, 32, 33]. During certain time frameworks and under given conditions, assessment of environmental impact is a very important systematic process. To measure the actual or potential impacts including psychosocial, physical, microbiological, and chemical hazard on the health case of humans or environment has a vital role [34–36]. After the obtained series of critical data from monitoring studies, quantitative environmental impact assessment (EIA) can be made. To provide better view for evaluating POP exposure and their adverse health effect on environment and human requires critical data obtained from the environment [37–39]. The EIA has several key stages, and it covers the risk level of all types of ecosystems. These stages are summarized in **Figure 2**. The EIA includes all activities which attempt to analyze and evaluate the effects of human stresses on natural and anthropogenic environments [36, 40–43].

4. Conclusion and future perspectives

The main aim of this study is to provide contemporary information on a variety of hydrocarbons present in the environment, exposure routes, and their adverse effects on ecosystem. Hydrocarbon sources, human health impact, and effect on the environment have been thoroughly investigated and presented. In light of this information, generated by natural or anthropogenic sources, hydrocarbons' muta-genic, teratogenic, and carcinogenic characteristics have caused serious concerns in today's environment; thus, various remediation techniques are needed to remove these hazardous chemicals from the environment. Therefore, some suggestions were presented as:

- All health authorities should develop standard methods for analysis of hydrocarbons and share it for all researchers.

- Researchers should develop more various remediation techniques available for hydrocarbons, and they should be applicable on every aspect of the environment such as soil, water, and air.

- After the treatment process, developed remediation techniques should not leave behind any second pollutant.

- Ecological risk assessment should be evaluated using the risk quotient.

- Techniques for removing hydrocarbons from the environment should be developed, but it is important that preventive measures can be taken to prevent these pollutants from entering the food chain and environment.

Author details

Muharrem Ince[1,3]* and Olcay Kaplan Ince[2,3]

1 Department of Chemistry and Chemical Processes, Tunceli Vocation School, Munzur University, Tunceli, Turkey

2 Faculty of Fine Arts, Department of Gastronomy and Culinary Arts, Munzur University, Tunceli, Turkey

3 Munzur University Rare Earth Elements Application and Research Center, Tunceli, Turkey

*Address all correspondence to: muharremince@munzur.edu.tr

IntechOpen

References

[1] Kamath R, Rentz JA, Schnoor JL, Alvarez PJJ. Phytoremediation of hydrocarbon-contaminated soils: Principles and applications. Studies in Surface Science and Catalysis. 2004;**151**:447-478

[2] Gkorezis P, Daghio M, Franzetti A, Van Hamme JD, Sillen W, Vangronsveld J. The interaction between plants and bacteria in the remediation of petroleum hydrocarbons: An environmental perspective. Frontiers in Microbiology. 2016;**7**:1836

[3] Gan S, Lau EV, Ng HK. Remediation of soils contaminated with polycyclic aromatic hydrocarbons (PAHs). Journal of Hazardous Materials. 2009;**172**(2-3):532-549

[4] Gitipour S, Ghasemi S, Shasemzade R. Methods for treatment of PAH contaminated soils; review and comparison. In: 4th International Conference on Energy, Environment and Sustainable Development. Jamshoro, Pakistan; 2016

[5] Connell DW. Basic Concepts of Environmental Chemistry. Boca Raton: CRC Press; 2005

[6] Ince M, Yaman M. High performance liquid chromatography-mass spectrometry for determination of benzo[a]pyrene in grilled meat foods. Asian Journal of Chemistry. 2012;**24**(8):3391-3395

[7] Ince M, Kaplan Ince O, Yaman M. Optimization of an analytical method for determination of pyrene in smoked meat products. Food Analytical Methods. 2017;**10**:2060-2067

[8] Kaplan Ince O, Ince M. Using box–Behnken design approach to investigate benzo[a]anthracene formation in smoked cattle meat samples and its' risk

assessment. Journal of Food Science and Technology. 2019;**56**:1287-1294

[9] Jin D, Jiang X, Jing X, Ou Z. Effects of concentration, head group, and structure of surfactants on the degradation of phenanthrene. Journal of Hazardous Materials. 2007;**144**(1-2):215-221

[10] Bansal V, Kim KH. Review of PAH contamination in food products and their health hazards. Environment International. 2015;**84**:26-38

[11] Ince M, Kaplan OI. An overview the toxicology of benzo(a)pyrene as biomarker for human health: A mini-review. Novel Techniques in Nutrition and Food Science. 2019;**4**(2):NTNF.000580.2019

[12] Douglas GA, Emsbo Mattingly S, Stout SA, Uhler AD, McCarthy KJ. Chemical finger printing methods. In: Murphy BL, Morrison RD, editors. Introduction to Environmental Forensics. 2nd edition. New York, NY: Academic; 2007. pp. 311-454

[13] Kim D, Kumfer BM, Anastasio C, Kennedy IM, Young TM. Environmental aging of polycyclic aromatic hydrocarbons on soot and its effect on source identification. Chemosphere. 2009;**76**(8):1075-1081

[14] Wang Z, Fingas M, Lambert P, Zeng G, Yang C, Hollebone B. Characterization and identification of the Detroit River mystery oil spill (2002). Journal of Chromatography. A. 2004;**1038**(1-2):201-214

[15] Zakaria MP, Takada H, Tsutsumi S, Ohno K, Yamada J, Kouno E, et al. Distribution of polycyclic aromatic hydrocarbons (PAHs) in rivers and estuaries in Malaysia: A widespread input of petrogenic PAHs. Environmental Science and Technology. 2002;**36**(9):1907-1918

5

[16] Stogiannidis E, Laane R. Source characterization of polycyclic aromatic hydrocarbons by using their molecular indices: An overview of possibilities. In: Whitacre D, editor. Reviews of Environmental Contamination and Toxicology (Continuation of Residue Reviews). Vol. 234. Springer, Cham; 2015

[17] Saber D, Mauro D, Sirivedhin T. Environmental forensics investigation in sediments near a former manufactured gas plant site. Environmental Forensics. 2006;7(1):65-75

[18] Hailwood M, King D, Leoz E, Maynard R, Menichini E, Moorcroft S, Pacyna J et al. Ambient Air Pollution by Polycyclic Aromatic Hydrocarbons PAH. Position Paper Annexes. 2001

[19] Wickramasinghe AP, Karunaratne DGGP, Sivakanesan R. PM10-bound polycyclic aromatic hydrocarbons: Concentrations, source characterization and estimating their risk in urban, suburban and rural areas in Kandy, Sri Lanka. Atmospheric Environment. 2011;45(16):2642-2650

[20] Osman KT. Soils, Principles, Properties and Management. Netherlands: Springer; 2013

[21] Ravindra K, Mittal AK, Grieken R. Health risk assessment of urban suspended particulate matter with special reference to polycyclic aromatic hydrocarbons: A review. Reviews on Environmental Health. 2001;16(3):169-190

[22] Boll ES, Christensen JH, Holm PE. Quantification and source identification of polycyclic aromatic hydrocarbons in sediment, soil, and water spinach from Hanoi, Vietnam. Journal of Environmental Monitoring. 2008;10(2):261-269

[23] Bakhtiari AR, Zakaria MP, Yaziz MI, Lajis MNH, Bi X, Rahim MCA. Vertical distribution and source identification of polycyclic aromatic hydrocarbons in anoxic sediment cores of Chini Lake, Malaysia: Perylene as indicator of land plant-derived hydrocarbons. Applied Geochemistry. 2009;24(9):1777-1787

[24] Tobiszewski M, Namieśnik J. PAH diagnostic ratios for the identification of pollution emission sources. Environmental Pollution. 2012;162:110-119

[25] Haffner D, Schecter A. Persistent organic pollutants (POPs): A primer for practicing clinicians. Current Environmental Health Reports. 2014;1:123-131

[26] Long M, Bonefeld-Jørgensen EC. Dioxin-like activity in environmental and human samples from Greenland and Denmark. Chemosphere. 2012;89:919-928

[27] Tavakoly Sany SB, Hashim R, Rezayi M, Salleh A, Rahman MA, Safari O, et al. Human health risk of polycyclic aromatic hydrocarbons from consumption of blood cockle and exposure to contaminated sediments and water along the Klang Strait, Malaysia. Marine Pollution Bulletin. 2014;84:268-279

[28] U.S.EPA. EPA's Reanalysis of Key Issues Related to Dioxin Toxicity and Response to NAS Comments. Vol. 1. Washington, DC; 2012

[29] Tavakoly Sany SB, Hashim R, Salleh A, Rezayi M, Mehdinia A, Safari O. Polycyclic aromatic hydrocarbons in coastal sediment of Klang Strait, Malaysia: Distribution pattern, risk assessment and sources. PLoS One. 2014;9(4):e94907

[30] WHO State of the science of endocrine disrupting chemicals 2012. United Nations Environment Programme and the World Health Organization. Geneva; 2013

[31] Ahmadzadeh S, Kassim A, Rezayi M, Rounaghi GH. Thermodynamic study of the complexation of p-isopropylcalix [6] arene with Cs^+ cation in dimethylsulfoxide-acetonitrile binary media. Molecules. 2011;**16**:8130-8142

[32] Liu K, Han W, Pan WP, Riley JT. Polycyclic aromatic hydrocarbon (PAH) emissions from a coal-fired pilot FBC system. Journal of Hazardous Materials. 2001;**84**(2-3):175-188

[33] Samanta SK, Singh OV, Jain RK. Polycyclic aromatic hydrocarbons: Environmental pollution and bioremediation. Trends in Biotechnology. 2002;**20**(6):243-248

[34] Rezayi M, Heng LY, Abdi MM, Noran NM, Esmaeili C. A thermodynamic study on the complex formation between tris (2-pyridyl) methylamine (tpm) with Fe^{+2}, Fe^{+3}, Cu^{+2} and Cr^{+3} cations in water, acetonitrile binary solutions using the conductometric method. International Journal of Electrochemical Science. 2013;**8**:6922-6932

[35] Saadati N, Abdullah MP, Zakaria Z, Tavakoly Sany SB, Rezayi M, Hassonizadeh H. Limit of detection and limit of quantification development procedures for organochlorine pesticides analysis in water and sediment matrices. Chemistry Central Journal. 2013;**7**:1-10

[36] Tavakoly Sany SB, Hashim R, Rezayi M, Salleh A, Safari O. A review of strategies to monitor water and sediment quality for a sustainability assessment of marine environment. Environmental Science and Pollution Research. 2014;**21**:813-821

[37] Law RJ, Bersuder P, Barry J, Deaville R, Reid RJ, Jepson PD. Chlorobiphenyls in the blubber of harbour porpoises (*Phocoena phocoena*) from the UK: Levels and trends 1991-2005. Marine Pollution Bulletin. 2010;**60**:470-473

[38] Rezayi M, Karazhian R, Abdollahi Y, Narimani L, Sany SBT, Ahmadzadeh S, et al. Titanium (III) cation selective electrode based on synthesized tris(2pyridyl) methylamine ionophore and its application in water samples. Scientific Reports. 2014;**4**:4664

[39] Tavakoly Sany SB, Salleh A, Sulaiman AH, Sasekumar A, Tehrani G, Rezayi M. Distribution characteristics and ecological risk of heavy metals in surface sediments of West Port, Malaysia. Environmental Protection Engineering. 2012;**38**:139-155

[40] Jazani RK, Tehrani GM, Hashim R. TPH-PAH contamination and benthic health in the surface sediments of Bandar-E-imam Khomeini-Northwest of the Persian Gulf. International Journal of Innovative Science, Engineering and Technology. 2013;**2**:213-225

[41] Tehrani GM, Sany SBT, Hashim R, Salleh A. Predictive environmental impact assessment of total petroleum hydrocarbons in petrochemical wastewater effluent and surface sediment. Environment and Earth Science. 2016;**75**:177

[42] Gitipour S, Sorial GA, Ghasemi S, Bazyari M. Treatment technologies for PAH-contaminated sites: A critical review. Environmental Monitoring and Assessment. 2018;**190**:546

[43] EHSC Environmental risk assessment. Environment, Health and Safety Committee [EHSC] of the Royal Society of Chemistry. 2008

Chapter 2

Source and Control of Hydrocarbon Pollution

Manish Srivastava, Anamika Srivastava, Anjali Yadav and Varun Rawat

Abstract

Hydrocarbon contamination is of great worry because of their widespread effect on all forms of life. Pollution caused by increasing the use of crude oil is ordinary because of its extensive application and its related transport and dumping problems. Crude oil contains a complex mixture of aliphatic, aromatic, and heterocyclic compounds. Soil naturally consists of heavy metals, and due to human action like refining of oil and use of pesticides, their concentration in soil is rising. Several areas have such high heavy metal and metalloid concentration that surrounding natural ecosystem has been badly affected. The reason is that heavy metals and metalloids limit microbe's activity rendering it unsuitable for hydrocarbon degradation, thus reducing its effectiveness. Environmental remediation is thus extremely necessary and involves with the elimination of pollutants from soil, air, and water. In the last several decades, different methods have been employed and applied for the cleanup of our environment which includes mechanical, chemical, and biochemical remediation methods. The hydrocarbon pollution consists of many aspects like oil spills, fossil fuels, organic pollutants like aromatics, etc. that are discussed below.

Keywords: aromatic hydrocarbons, organic and inorganic pollutants, bioremediation, chemical remediation

1. Introduction

Contamination of hydrocarbon occurs due to toxic organic substances, petroleum, and pesticides which is a serious concern for the environment. Contamination caused by petroleum hydrocarbon is a matter of worry because these are harmful for various life forms. Crude oil contamination is common due to its extensive use and its related dumping process and accidental spills. Complex mixture of a large range of high and low molecular weight hydrocarbons makes up the petroleum. The complex mixture of petroleum consists of saturated and branched alkanes, alkenes, and homo- and heterocyclic naphthenes; aromatics consisting of heteroatoms such as heavy metal complexes and N, S, and O; hydrocarbon consisting of different functional groups such as ethers, carboxylic acids, etc.; and large aromatic molecules such as asphaltenes, resins, and naptheno-aromatics.

Heavy metals are present in crude oil, and its heavy metal content is associated with porphyrins which is the pyrrolic structure. Lube oil waxes, light oil, asphaltenes, naphtha, diesel, kerosene, etc. are the several fractions in which the petroleum is refined. Light ends is the term that is used for the light fractions which are distilled at atmospheric pressure, and heavy ends is used for heavy fractions such as asphaltenes and lube oil. Due to different hydrocarbon compositions of light and heavy ends of petroleum, light ends consists of a lower percentage of aromatic compounds and lower molecular weight saturated and unsaturated hydrocarbons, while heavier ends consists of higher molecular weight saturated and unsaturated hydrocarbons, aromatic compounds with high molecular weight, and organometallic compounds. This part is relatively affluent in metals and nitrogen, sulfur, and oxygen-containing compounds [1].

Concentration of heavy metal is rising in the soil as a consequence of human action. There is a large impact of higher heavy metal and metalloid concentration in some areas [2].

2. Hydrocarbon pollution

This is caused mainly by accidents on oil platforms and ships used for hydrocarbon transportation but also by discharging water into the sea which is used to wash tanks of tanker vessels. Crude oil and petroleum products form a waterproof film on water that prevents the oxygen exchange between environment and water causing damages to plants, animals, and human beings. Nowadays during transport overseas, "double-hulled" tankers are used to avoid leaks in case of accidents. Best international practices are adopted with regard to oil platforms to face or eventually adequately deal with any type of inconvenience.

3. Organic pollutants

With the onset of industrialization, the use and buildup of organic compounds have increased. Major sources which are responsible for organic contaminants are anthropogenic activities including the use of fuels, solvents, and pesticides. Various organic compounds are harmful and are related to health concerns globally.

Diverse sources are responsible for the generation of hydrocarbons in sediments which are categorized below [3, 4]:

- Anthropogenic sources

- Petroleum inputs

- Partial burning of fuels

- Fires of forest and grass

- Biosynthesis of hydrocarbons by marine or terrestrial organisms

- Diffusing from the petroleum source rocks, reservoirs, or mantle

Organic pollutant is responsible for environmental and health-related problems; hence bioremediation provides an efficient explanation to this problem [5].

3.1 Polycyclic aromatic hydrocarbons (PAHs)

PAHs are considered to be ubiquitous contaminants. There are 100 diverse compounds of polycyclic aromatic hydrocarbons present. PAHs are seldom used for the industrial purpose, but only few are used for the manufacturing of pesticides, dyes, and plastics and for the production of medicines. Polycyclic aromatic hydrocarbons are produced on partial burning of organic matters [5]. PAHs due to carcinogenic and mutagenic nature are highly poisonous to organisms. The degradation of PAHs is predominantly slow with high molecular weights because due to low hydrophobicity and water solubility it has a tendency to accumulate in sediments [6]. PAHs have been classified as a priority pollutant by the USEPA which has classified 16 individual PAHs as pollutants due to its poisonous, carcinogenic, and mutagenic nature [7].

3.2 Polychlorinated biphenyls (PCBs)

Polychlorinated biphenyls (PCBs) due to carcinogenicity, toxicity, and slow biodegradation in the nature are well thought-out to be the worst pollutants [8] of commercial PCBs of about hundreds of thousands of metric tons are persevere in aquatic sediments [9]. In adhesives and lubricants, dielectric fluids in flame retardants, transformers, hydraulic fluids, and plasticizers, PCBs are widely used. PCBs are released from disposal and spillage [10].

3.3 Polychlorinated dibenzo-p-dioxins and dibenzofurans (PCDD/Fs)

Polychlorinated dibenzo-p-dioxins and dibenzofurans (PCDD/Fs) are still present in deep sediment layers which are deposited decades ago. Toward biotic and abiotic degradation processes, PCDD/Fs are often well-thought-out to be recalcitrant [11]. PCDD/Fs are the most notorious pollutants present in nature [12].

Table 1 Microorganisms studied.

Pollutants	Organisms	Function	References
2,4,6-Trinitrotoluene (TNT)	*Methanococcus* sp.	Biotransformation	Boopathy and Kulpa [13]
Atrazine	*Pseudomonas* sp. (ADP)	Biodegradation	Newcombe and Crowley [14]
Chlorpyrifos	*Enterobacter* strain B-14	Biodegradation	Singh et al. [15]
Dibenzothiophene (DBT)	*Rhizobium meliloti*	Biodegradation	Frassinetti et al. [16]
Hexahydro-1,3,5-trinitro-1,3,5-triazine (RDX)	*Acetobacterium paludosum*	Biodegradation	Sherburne et al. [17]
PAHs	*Clostridium acetobutylicum*	Biodegradation	Zhang and Hughes [18]
Phenanthrene, PAH	*Pseudomonas* sp., *Pycnoporus sanguineus, Coriolus versicolor, Pleurotus ostreatus, Fomitopsis palustris, Daedalea elegans*	Biodegradation	Arun et al. [19]
Polychlorinated biphenyl (PCB)	*Agrobacterium, Bacillus, Burkholderia, Pseudomonas,* and *Sphingomonas*	Biodegradation	Aitken et al. [20]
Polycyclic aromatic hydrocarbon (PAH)	*Rhodococcus erythropolis* TA421	Biodegradation	Chung et al. [21]
	Rhizobium sp.		Damaj and Ahmad [22]
	Fungi		Atagana [23]

Table 1.
Microorganisms studied or bioremediation function.

4. Inorganic pollutants

Human sources are mainly responsible for the heavy metal contamination, but contamination due to natural and biological processes are also common which includes:

1. Mineral weathering over time.

2. Erosion and volcanic actions.

3. Forest fires and biogenic resource.

4. Vegetation causes release of particles.

Cellular binding sites of microbes are responsible for the absorption of heavy metals. By various mechanisms, heavy metals can be complexed with extracellular polymers of microbes. Organic contaminants can be mineralized by these microorganisms and convert into metabolic intermediates which can be utilized as primary substrates for growth of the cell. Heavy metals can be eliminated from the metal-polluted soil by microbes which can change the heavy metal oxidation state by immobilizing them [24]. Research on bioremediation of heavy metals by microbes has not been carried out extensively due to metal adsorption and incomplete knowledge of the genetics of the microbes.

5. Sources and effects of hydrocarbon-contaminated wastewater effluents

Numerous sources such as pesticides, petroleum, or different harmful organic substances which are discharged into the water streams as effluents are responsible for the hydrocarbon pollution into the wastewater. Water contaminated with hydrocarbons is known to be carcinogenic, neurotoxin, and mutagenic to flora and fauna [25]. Contaminated lands, oil spillage, pesticides, automobile oils, and urban stormwater discharges are the major causes for the hydrocarbon contamination.

Oil spill is one of the major sources of hydrocarbon contamination. Oil spills caused mainly by accidents on oil platforms and ships are needed for transportation of hydrocarbon but also by disposal of water into the sea which is used to wash tanks of tanker vessels [2]. Underground oil storage tanks and leaking pipelines are also responsible for oil spilling in water [26, 27].

Increase use of vehicles and automobiles leads to increase in utilization of automobile oil, which is the major cause of hydrocarbon contamination in water. This type of contamination occurs when oil from the car drops onto the ground and leaks; it could be washed into water streams by runoffs [28].

Pesticides are another source of hydrocarbon contamination in water. Pesticides include herbicides, fungicides, and insecticides. Only small amount of pesticide is able to achieve the target, while the major proportion stays in the soil, and it can be washed away by the rain in the water stream [29]. Herbicides, out of all the pesticides, are most hazardous because it is directly applied on the soil in order to kill the weed and can be washed away during rainfall into the water streams.

Another source of hydrocarbon contamination in water is the land where some type of industrial action is being carried out. These lands contaminated by hydrocarbons or toxic organic compounds are washed due to rainfall into the water steams, thus causing pollution [30].

One of the main sources of hydrocarbon pollution is the discharge of urban stormwater. In urban communities, car parks and roads are frequently polluted by gasoline and oil from the vehicles, and during rainfall, these pollutants are washed into water streams and hence can contaminate them [30, 31].

Wastewater contaminated by hydrocarbons has an adverse effect in nature, animals, human beings, and plants. Lack of oxygen, decrease in crop yield, and effects on aquatic plants are various effects of hydrocarbon contamination in nature. There would be decrease in the crop yield and available food for household due to inappropriate crop's growth when the farmland is irrigated by water contaminated with hydrocarbon [32, 33]. Soil fertility can be decreased to an extent due to the presence of oil in water due to the reason that most of the vital nutrients are no longer accessible for crop consumption which results in the decrease of the crop yield. The reduction in the yield of crop results in the decrease of the farmer's earnings [34, 35].

Oxygen shortage is another environmental effect of hydrocarbon contamination. The main source of oxygen in nature is the economic trees which rely on rainfall or on the water steams for their growth. Oil spills can inhibit root penetration due to hydrocarbons which can block the pores of the soil, thereby removing water and air [36]. This results in the death of such plant or distortion in the growth and hence causes oxygen shortage for human utilization [37]. Hydrocarbon contamination in water avoids the penetration of light into the water and the exchange of gases for consumption by aquatic plants. This leads to the death of the plant because plant becomes incapable to photosynthesize and hence can affect the food chain. Plants consume the pollutants from the contaminated water which can be passed to humans and animals through the food chain [38].

Polycyclic hydrocarbons are toxic and found to have serious effects on human beings. The immune system, liver, respiratory system, reproductive system, circulatory system, kidney, etc. are the organs which are affected due to the hydrocarbon ingestion [33]. Individual's susceptibility and level of exposure are the factors on which the degree of damage depends [2]. Cancer risk and hormonal problems that can disturb developmental and reproductive processes are the other effects of effluents polluted by hydrocarbons on human beings [39–41].

Discharge of wastewater contaminated with hydrocarbon into the water streams poses risk to animals through absorption, breathing, and ingestion. Sea birds are the most exposed to the hydrocarbon pollutant because it spends majority of its time near the water bodies [42]. There is unusual decrease in the temperature due to the destruction of the protective layer of the feathers in sea birds as a result of the presence of oil in water [43]. Scavengers such as ravens and vultures are also in danger when they consume preys and contaminated fish [44]. Water contaminated with hydrocarbon is consumed through gills of the fish during the respiration and accumulates in the gall bladder, liver, and stomach, and thus the fish becomes unhealthy for human utilization [45].

6. Remediation

Polluted land or water systems have become a serious concern for human health. Over the past few decades, several methods have been developed and applied for the cleanups. The degradation either biological or chemical of petroleum which is a complex mixture of chemical substances is difficult because different treatments are required for different classes of compounds. Hence, remediation of oil-containing environment is not easy. Remediation strategies are decided after knowing the oil composition and physicochemical nature of the polluted site. Physical and chemical properties and pH of the polluted water/soil are the different factors on which the crude oil degradation depends. Oil-producing wells are generally situated near seashore, so due to this reason, water is contaminated mostly by oil spills during oil production operation. Oil spills are controlled by biological and chemical methods. Out of these two methods, chemical method is more frequently used. Bioremediation is gaining worldwide attention.

7. Remediation techniques for hydrocarbons

Contamination due to petroleum is widespread in the environment and contaminates surface and groundwater [46]. Several operations in petroleum exploration, leaking of underground storage tanks, and its production and transportation are responsible for affecting the environment [47]. Contamination causes threat to human health and safety and can affect nature by contaminating surface and groundwater [46].

Efforts are made both nationally and internationally in order to remediate the pollution caused by hydrocarbon contamination which can cause environmental and health risk. There are three methods involved in the remediation of sites contaminated due to hydrocarbon [2, 48]:

1. Phytoremediation

2. Bioremediation

3. Chemical remediation

7.1 Phytoremediation

Phytoremediation is the process which involves the use of plants for the degradation, extraction, and elimination of the contaminants from the air, water, and soil [40, 49–51]. It includes various mechanisms which can lead to degradation of contaminants, dissipation, immobilization, and accumulation [52, 53]. Various phytoremediation applications with examples are systematically given in **Table 2**.

7.1.1 Mechanisms of phytoremediation

Contaminated land and water are remediated more feasibly by using plants involving a variety of pollutant attenuation mechanisms than physical and chemical remediation techniques [54–58]. Plants due to their sedentary nature had developed various abilities for dealing with hazardous compounds. Plants serve as solar-driven pumping and filtering systems as they take up pollutants from the soil through the roots which is transported to various parts of the plant by the help of plant tissues where they can be volatilized, metabolized, or sequestered [57, 59]. Different types of mechanisms are used by the plant for removing the pollutants from the soil. They consist of biophysical and biochemical processes such as adsorption, translocation, and transport, as well as mineralization and transformation by plant enzymes are the mechanisms of phytoremediation [8]. Halogenated substances like TCE are degraded by plants using oxidative degradation pathways, and it includes plant-specific dehalogenases. After the death of the plant, the dehalogenase activity is still maintained [60]. Laccases, P450 monooxygenases, nitroreductases, dioxygenases, phosphatases, peroxidases, dehalogenases, and nitrilases are various contaminant-degrading enzymes which are present in plants [61–63]. The basic physiological mechanisms involved in phytoremediation in higher plants and related microorganisms, such as

Application	Media	Contaminants	Typical plants
Phytotransformation	Soil, groundwater, landfill leachate, land application of wastewater	Herbicides, aromatics, chlorinated aliphatics, nutrients, ammunition waste	Phreatophyte trees (popular, willow, cottonwood, aspen) Grasses (rye, Bermuda, sorghum, fescue) Legumes (clover, alfalfa, cowpeas)
Rhizosphere bioremediation	Soil, sediments, land application of wastewater	Organic contaminants (pesticides, aromatics, and polynuclear aromatic hydrocarbons)	Phenolic releasers (mulberry, apple, Osage orange) Grasses with fibrous roots (rye, fescue, Bermuda) for contaminants 0.3 ft deep Phreatophyte trees for 0.10 ft Aquatic plants for sediments
Phytostabilization	Soil, sediments	Metals (Pb, Cd, Zn, As, Cu, Cr, Se, U), hydrophobic organics (PAHs, PCNBs, dioxins, furans, pentachlorophenol, DDT, dieldrin)	Phreatophyte trees to transpire large amounts of water for hydraulic control Grasses with fibrous roots to stabilize soil erosion Dense root systems are needed to sorb/bind contaminants
Phytoextraction	Soil, brown fields, sediments	Metals (Pb, Cd, Zn, Ni, Cu) with EDTA addition for Pb selenium (volatilization)	Sunflowers Indian mustard Rape seed plants Barley Hops Crucifers Serpentine plants Nettles Dandelions
Rhizofiltration	Groundwater, water and wastewater in lagoons or created wetlands	Metals (Pb, Cd, Zn, Ni, Cu), radionuclides (137Cs, 90 Sr, U), hydrophobic organics	Aquatic plants: emergents (bulrush, cattail, coontail, pondweed, arrowroot, duckweed); submergents (algae, stonewort, parrot's feather, Eurasian watermilfoil, hydrilla)

Table 2.
Application of phytoremediation with examples.

mineral nutrition, photosynthesis, transpiration, and metabolism. The root of the plant is responsible for the uptake of the organic and inorganic compounds from the soil, and it can bind and stabilize substance on its external surfaces on interaction with microorganism in the rhizosphere. Uptake or release of molecules occurs through exchanging gases from the aerial plant's parts with the atmosphere [64]. For addressing different contaminants in different substrates, six phytotechnologies have been recognized by Interstate Technology and Regulatory Cooperation:

1. For organic contaminants, phytotransformation is ideal in all substrates.

2. Rhizosphere bioremediation is used in soil containing organic contaminants.

3. Phytostabilization is used in soil for organic and inorganic pollutants.

4. Phytoextraction is useful in substrates containing inorganic pollutants.

5. Phytovolatilization is used for volatile substances.

6. Hydraulic flow can be controlled in the contaminated environment by using evapotranspiration.

7.2 Bioremediation

Bioremediation is a cost-efficient method used for the treatment of soil polluted with oil and wastes of petroleum consisting of biodegradable hydrocarbons and indigenous microbes.

The management of suitable levels of nutrient fertilizer addition, moisture control to optimize soil degradation by microorganisms, aeration and mixing, and pH amendment are required for the process of land treatment [65].

Enzymes attack on some inorganic compounds and on most of the organic compounds through the activities of living organisms. Bioremediation is the technique which involves the productive use of the biodegradative process for the elimination or detoxification of pollutants from the environment.

Oil spill causes contamination of soil which is considered as the chief worldwide concern. Pollution of soil due to petroleum causes a serious effect to human being, affects the groundwater, decreases the agricultural production of the soil, and causes economic loss and ecological problems. Plants, animals, microorganisms, and humans are affected by the toxicity of the petroleum hydrocarbons. Oil spill and accidents occur due to the transportation of crude oil which is generally through tankers on water or through land pipeline. Problems of the oil contamination occur mostly due to the reason that the main oil-producing countries are not the chief oil clients; hence petroleum is transported to the consumption area. Certain microorganisms are accountable for the petroleum hydrocarbon degradation and are used as the resource of carbon and energy for growth and maintenance. Soil contamination can be remediated by many ways including both physicochemical and biological techniques.

Biological techniques are more economical and proficient than physicochemical techniques. The degradation rate of petroleum products is increased by developing several remediation methods. Bioremediation through microorganism is considered to be the most effective method in comparison to other biological methods, but the high molecular weight hydrocarbons with low adsorption and solubility limit their accessibility to microorganisms.

7.2.1 Principle of bioremediation

Composite mixture of diverse chemical substances makes up the crude oil. Oil and its component are recognized by microbes using bioemulsifiers and biosurfactants, and then they join themselves; hydrocarbon is used as the resource of carbon and energy. High molecular weight hydrocarbons due to their low adsorption and solubility limit their accessibility to microorganisms. Oil biodegradation rates are improved by the biosurfactant's addition which increases the elimination and solubility of these pollutants.

The oil constituents vary particularly in susceptibility, volatility, and volubility to biodegradation. A number of substances are easily degraded, some are non-biodegradable, and some oppose degradation. Diverse species of microbes preferentially attack diverse compounds due to this biodegradation of petroleum that occurs at different rates but concurrently. Enzymes produced by microorganisms in the presence of sources of carbon are accountable for attacking the hydrocarbon molecules. Hydrocarbon present in the petroleum is degraded by different enzymes and metabolic pathways. Hydrocarbon degradation is prevented by the lack of suitable enzyme [66].

Bioremediation process involves the utilization of natural microorganisms for the decontamination of atmosphere [67]. This process converts pollutants into useful or nontoxic substances by using bacteria, fungi, and yeast which are the naturally occurring microorganisms [40]. This is also a process in which microorganisms restore the quality of the environment by degrading and metabolizing the chemical substances [48]. **Table 3** represents the main microorganisms which are included in the remediation of hydrocarbons.

7.2.2 Microorganisms

7.2.2.1 Bacteria

Microbial species has efficient hydrocarbon degradation capability in natural environments. Various microbial species have been isolated from heavily polluted coastal areas, variety of oil spill, or soil contaminated by petroleum. These are isolated on the basis of their capability to metabolize different sources of carbon such as aliphatic and aromatic compounds and their chlorinated derivate. Enrichment culture procedures were used for obtaining the microorganisms, and for the selection criterion, maximum final cell concentration or maximum specific growth rate was used. Various microorganisms such as fungi, microalgae, bacteria, and yeast [68] are used for degrading the petroleum hydrocarbons. Out of these microorganisms, bacteria play a significant role for hydrocarbon degradation. Rapid degradation of low molecular weight alkanes is reported by various studies. The capability of microorganisms to use hydrocarbons to assure the growth of cell and energy requirements by degrading hydrocarbon is the driving force for the petroleum biodegradation. Biodegradation of petroleum is carried out more extensively by mixed cultures in comparison to pure culture [69]. Adequate indigenous microbial community in many ecosystems is capable of biodegradation of oil, but for oil degradation metabolic activity, environmental conditions should be favorable. Indigenous microorganisms have several advantages than adding microorganisms for hydrocarbon degradation.

7.2.2.2 Fungi

For the biodegradation of hydrocarbons in soils, fungi play a more vital role than bacteria. Filamentous fungi which are found in aquatic structures are mostly related with surface films and sediments. The enzymatic processes used by mammalian organizations are also used by fungi in polycyclic aromatic hydrocarbons (PAHs).

Bacteria	Yeast and fungi
Achromobacter	Aspergillus
Acinetobacter	Candida
Alcaligenes	Cladosporium
Arthrobacter	Penicillium
Bacillus	Rhodotorula
Brevibacterium	Sporobolomyces
Corynebacterium	Trichoderma
Flavobacterium	Fusarium
Nocardia Pseudomonas Vibrio	Trichoderma

Table 3.
List of microorganisms for bioremediation.

Two major types of cytochrome P450 monooxygenases have been well characterized in yeasts and filamentous fungi. Several fungi have the ability to oxidize polycyclic aromatic hydrocarbons to phenols, dihydrodiols, and other metabolites and conjugates, but only some fungi such as *Phanerochaete chrysosporium* have the capability to catabolize them totally to CO_2.

Example:

i. *Mitosporic Ascomycota*

ii. *Dothiorella Aureobasidium*

iii. *Saccharomycetales candida*

7.2.2.3 Yeast

The biodegradability of various yeasts decreases from n-alkanes > branched alkanes > low molecular weight aromatic hydrocarbons > cycloalkanes > high molecular weight aromatic and polar compounds.

Bioremediation process involves the detoxification of pollutants due to the various metabolic capabilities of microorganisms which is the developing method for elimination of contaminants from nature together with the yields of the petroleum industry [70]. Bioremediation technique is considered to be cost-effective and noninvasive. Petroleum and other hydrocarbon contaminants can be eliminated from the atmosphere by using microorganisms which is considered as primary mechanism, and it is the cheaper method in comparison to other remediation technologies. Microorganisms having suitable metabolic capabilities are the essential requirement.

Alkylaromatic degradation is carried out by various microorganisms such as *Arthrobacter*, *Mycobacterium*, *Sphingomonas*, *Burkholderia*, *Rhodococcus*, and *Pseudomonas*.

Fungi, bacteria, and yeast are accountable for the biodegradation of hydrocarbons in the environment. Six percent [71] to 82% [72] is the reported efficiency of biodegradation for soil fungi, 0.003–100% [73] for marine bacteria, and 0.13% [71] to 50% [72] for soil bacteria. Complex mixtures of hydrocarbons such as crude oil in freshwater, aquatic environments, and soil are degraded by mixed populations with overall wide enzymatic capacities [74].

Bioremediation involves two processes as follows:

1. Bioaugmentation

2. Biostimulation

7.2.2.3.1 Bioaugmentation

Bioaugmentation process involves the degradation of the harmful hydrocarbons by the addition of microorganisms in order to achieve the pollutant reduction [67]. It is also the injection of polluted water with microorganisms capable of hydrocarbon degradation [48]. This process sometimes involves biodegradation of the hydrocarbon pollutants by adding the genetically engineered microorganisms into the polluted water [75]. Bioaugmentation process is not often used for the hydrocarbon degradation because microorganisms responsible for hydrocarbon degradation naturally exist in the environment. Bioaugmentation process is not so much effective to be used in oil spill remediation sites, and nonindigenous microorganisms used in this process can cause competition with the microbes already present in the environment [76].

7.2.2.3.2 Biostimulation

Biostimulation is the process which involves degradation of the harmful compounds by adding the nutrients required by indigenous hydrocarbon-degrading microbes [67]. The growth of microorganisms responsible for the degradation of oil during oil spillage is activated by the increase in carbon. The tendency of the microorganisms to degrade the hydrocarbons is enhanced by addition of suitable concentration of supplemental nutrients. Due to this reason, microorganisms are competent of achieving their utmost rate of growth and consequently the utmost rate of contaminant uptake [77, 78]. The maximum biostimulation is achieved by obtaining the ideal nutrient concentration which is required for the utmost growth of the microorganisms and maintaining concentration as long as possible for microorganisms [79].

7.3 Chemical remediation

This process requires the use of chemicals. Contaminants can be treated by using various chemicals. Chemicals usually have the capability of altering the contaminant's chemical and physical properties [80]. Dispersants, solidifiers, and chemical oxidants are the three categories in which the chemical remediations are grouped [2, 48, 52].

7.3.1 Dispersants

Slick of oil can be broken down into smaller droplets by surfactants which are present in dispersants, and these droplets undergo rapid dilution by transferring it into the water and can be effortlessly degraded [81]. Chemical dispersants can raise the oil droplet surface area which results in an increased rate of natural biodegradation, and this process makes the oil less sticky to the surface by slowing down the development of oil-water emulsions and allows fast treatment [82]. This method makes oil spills less harmful for living organisms and the marine life. This is achieved by converting oil slicks into droplets which in turn can be degraded by bacteria [2, 81]. Nokomis 3-F4, Slickgone NS, Finasol OSR 52, SPC 1000™, Neon AB3000, ZI-400, Corexit 9500, Corexit 8667, and Saf-Ron Gold are some of the examples of chemical dispersants [83].

7.3.2 Solidifiers

In this method oil is removed by physical method which involves the interaction of dry granular materials with the oil and converts its liquid state into rubberlike solid state. Dry particulate and semisolid substances such as balls, pucks, sponge, etc. are the various forms in which the solidifiers can be applied. Solidification can be enhanced by using the solidifiers in the seas because mixing energy is provided by the seawater. Solidifiers are difficult to recover after solidification, and it is less efficient, which are the major drawbacks for the use of the solidifiers [82, 84].

7.3.3 Chemical oxidation

This technique involves the usage of chemical agents which are capable of oxidizing the organic pollutants [85]. These chemical agents are introduced by the help

Chemical treatment	Advantages	Disadvantages
Dispersants	Suitable in all weather condition and for wide range of oils	No oil recovery Not effective on highly viscous, non-spreading, and waxy oil
	Accelerates by degradation of the oil by natural processes Advanced formulations have reduced the previous concerns about toxicity Less man power needed Less expensive than mechanical methods	The localized and temporary increase in the amount of oil in water concentration that would have an effect on the surrounding marine life If dispersion is not achieved, other response method effectiveness may reduce on less disperse oil
Solidifiers	All weather conditions Quick	Lack of practical application Large amount required Selected oil Not effective No oil recovery

Table 4.
Advantages and disadvantages of chemical treatment.

of the mixing apparatus and injection in water or soil at the contaminated site. The usefulness of the process is found to depend upon oxidant quality, efficient contact between pollutant and oxidant, geological conditions, and oxidant's residence time [86]. This process is rapid and can be applied in all weather situations which are some of the advantages of this process. **Table 4** represents the details of other advantages and disadvantages.

8. Chemical and mechanical remediation methods

8.1 Oil spilled on the sea surface

There are various techniques involved for the elimination of oil from the surface of the sea and to avoid the oil to reach the shoreline. The widely used methods are mechanical recovery and the application of dispersants. The crude oil spreads over the sea surface because it is lighter than water and the thickness of the oil film becomes very thin in a small time. Type of oil, temperature of atmosphere, tide, temperature of water, and wind are the factors on which the velocity of oil spreading depends.

If oil spills accidentally, then the spreading of the oil can be prevented by using skimmers and booms which can control the spill to a short area, and finally the oil can be collected into the container. Oil can be solubilized by applying biosurfactants which are generally not detrimental to nature.

For oil spill remediation, at times in situ oil burning is also used as an optional method; but in situ method is useful only when the spilled layer of oil is floating on the surface of the water, oil spill is fresh, or after the oil has been converted into a smaller area by the booms. The above technique has some drawback that aquatic system gets polluted by the by-products and smoke generated as a result of burning of oil. Weather, tides, and ocean currents are the factors on which the usefulness of the cleaning method depends. If the oil reaches the shoreline, different methods are applied to clean up the gravels and sand. Oil is absorbed sometimes by oil sorbents similar to sponge. Oil is removed from the oiled vegetation by washing with water, but the plants damaged severely should be detached completely.

When the amount of contaminated water is less than ex situ, remediation method is applied by pumping the contaminated water to the processing site. The shore sand and gravels are removed and cleaned in a different place from the contaminated site.

8.2 Oil spilled on soil

Pollution of soil occurs due to leakages from pipes and wellheads during offshore oil production and drilling operations, leakage from underground storage tanks of petroleum, overflow from gathering stations, petroleum yields, and inappropriate dumping of waste of petroleum. During the excavation, transport, and handling of polluted material, significant risk may be created by this method. For the final disposal of the substance, it is very hard to locate new landfill sites. There is continuous requirement of monitoring and maintenance of separation barriers since the pollutant remains on the site, and hence cap and containment technique is the temporary solution.

Methods for the treatment of soil contamination are as follows:

1. In situ

2. Ex situ

8.2.1 In situ method

This method involves physicochemical processes including air sparging, soil air extraction, or by combinations of these two methods applied to the soil at the contaminated site. Vertical & horizontal fossil fuel drilling equipment's are used *in-situ* treatment. This technique is more efficient on sandy soil than on clay soils. Soil pollutant can be taken out by using air sparging which is also known as soil venting.

The growth of aerobic bacteria on oxygen feeding is accelerated by the help of this method. Air sparging can be also performed under the water table if the contamination takes place in the groundwater through extraction wells or to the surface by gravity segregation. The oil can be extracted from the oil saturated ground water or partially saturated soil by using a process called as slurping.

The volatile components which are trapped in the soil are extracted by injecting steam into the contaminated soil.

8.2.2 Ex situ

This technique involves the elimination and transportation of polluted soil to off-site remediation ability. Various processes are used to perform the ex situ remediation which is as follows:

Land farming process is used in which soil polluted with oil is excavated and spread above a bed where it once in a while is tilled until the contaminants are degraded. Fifteen to 35 cm of soil surface is treated with the help of this technique. Composting involves the increase in the development of the microbial species by mixing polluted soil with harmless organic compounds to contaminated soil. Bioreactors are used for the bioprocessing of polluted soil, sediment, and water in which the three phases, gas, soil, and liquid, are mixed continuously in order to enhance the biodegradation rate. Before loading the contaminated soil to the bioreactors, the soil is pretreated. Contaminants undergo chemical reaction and convert harmful compounds into nontoxic compounds. Dechlorination or UV is used for the catalyzation of the oxidation reactions. These techniques have a few limitations such as high cost due to the complication of the method required, while bioremediation due to natural biological action is a choice which provides the chance to degrade the hydrocarbon contaminants.

9. Application of bioremediation

1. Ecologically sound, natural process; there is an increase in the number of the existing microorganisms when the contaminants are present, and the microbial population decreases naturally when the contaminants are degraded. The residues such as water, carbon dioxide, and fatty acids obtained as a result of the biological treatment are usually nonhazardous product, and the obtained CO_2 can be used for the photosynthesis process by the plants.

2. Bioremediation is responsible for destroying the target chemicals in place of transferring the contaminants from one place to another.

3. Other techniques which are used for the cleanup of harmful waste are more costly than bioremediation. For example, through the cleanup of the Exxon Valdez spill, the cost of 1-day physical washing is more than bioremediating 120 km of shoreline.

4. Bioremediation deals with in situ treatment and does not involve the transfer of a large amount of the polluted wastes off-site, and the risk due to the transportation can be overcome.

5. Microbe efficiency can be enhanced by using nutrient formulation in the bioremediation process.

6. The residues such as CO_2, fatty acids, water, etc. obtained from the biological treatment are generally nonhazardous.

It is a less costly technique than other techniques which are used for cleaning up of the toxic waste.

Hydrocarbons due to their different solubility from polar compounds such as methanol have lower polarity and hence have low solubility. Degradation of hydrocarbons is not only determined by solubilization. Many microorganisms are responsible for increasing the surface area of the substrate by excreting emulsifiers including *Bacillus licheniformis*, *Pseudomonas putida*, *Bacillus cereus*, *Pseudomonas aeruginosa*, *Bacillus subtilis*, and *Bacillus laterosporus*. Absorption of hydrophobic substance is facilitated by change in the cell surface by microorganisms. The behavior of individual hydrocarbons as well as mixtures can be changed by changing the physicochemical character of hydrocarbons [74].

10. Conclusion

Hydrocarbon pollutants have a widely applicable consequence on land, aquatic, as well as atmospheric ecosystem. This has been a problem ever since the use of fossil fuels and industrial revolution started. The unparalleled growth in populations with frequent oil spills, leakages in pipelines, and rampant use of pesticides contribute to substantial increase in pollution. These together are threatening the lives of animals and native microbiological population in land, air, and water surfaces and subsurfaces. Thus environmental remediation is the most important aspect of human survival. This book not only highlights the causes but also explains the techniques used in pollution rectifications. The various remediations described in this chapter are (i) phytoremediation, (ii) bioremediation, and (iii) chemical remediation.

Author details

Manish Srivastava[1*], Anamika Srivastava[1], Anjali Yadav[1] and Varun Rawat[2]

1 Banasthali Vidyapith, Rajasthan, India

2 Amity University, Gurugram, Haryana, India

*Address all correspondence to: dr.srivastava2480@gmail.com

IntechOpen

References

[1] Available from: http://www.
dep.state.fl.us/waste/quick_topics/
publications/pss/pcp/Petroleum
ProductDescriptions.pdf

[2] Abha S, Singh CS. Hydrocarbon
pollution: Effects on living organisms,
remediation of contaminated
environments and effects of heavy
metals co-contamination on
bioremediation. In: Romero-Zeron
L, editor. Introduction to Enhanced
Oil on Recovery (EOR) Processes and
Bioremediation of Oil Contaminated
Sites. China: InTech Publisher; 2012.
pp. 186-206. ISBN: 978-953-51-0629-6

[3] Readman JW, Fillmann G, Tolosa
I, Bartocci J, Villeneuve JP, Catinni
C, et al. PAH contamination of the
Black Sea. Marine Pollution Bulletin.
2002;**44**:48-62. DOI: 10.1016/
s0025-326X(01)00189-8

[4] Kim GB, Maruya KA, Lee RF, Lee
JH, Koh CH, Tanabe S. Distribution
and sources of polycyclic aromatic
hydrocarbons in sediments from
Gyeonggi Bay, Korea. Marine Pollution
Bulletin. 1999;**38**:7-15. DOI: 10.1016/
s0025-326X(99)80006-X

[5] US-EPA Great Lakes National
Program Office. Realizing Remediation:
A Summary of Contaminated Sediment
Remediation Activities in the Great
Lakes Basin; 1998

[6] Readman JW, Mantoura RFC, Rhead
MM, Brown L. Aquatic distribution and
heterotrophic degradation of polycyclic
aromatic hydrocarbons (PAH) in the
Tamar estuary. Estuarine, Coastal and
Shelf Science. 1982;**14**:369-389. DOI:
10.1016/s0272-7714(82)80009-7

[7] IARC (International Agency for
Research on Cancer). IARC Monographs
on the Evaluation of the Carcinogenic
Risk of Chemicals to Humans:
Polynuclear Aromatic Compounds Part
I. Lyon: IARC Press; 1983

[8] Meagher RB. Phytoremediation
of toxic elemental and organic
pollutants. Current Opinion in Plant
Biology. 2000;**3**:153-162. DOI: 10.1016/
s1369-5266(99)00054-0

[9] NRC National Research Council.
Polychlorinated Biphenyls; 1979

[10] Scragg A. Bioremediation.
Environmental Biotechnology.
2005:173-229

[11] Uchimiya M, Masunaga S. Time
trend in sources and dechlorination
pathways of dioxins in agrochemically
contaminated sediments. Environmental
Science and Technology. 2007;**41**:
2703-2710. DOI: 10.1021/es0627444

[12] Kaiser J. Just how bad is dioxin?
Science. 2000;**288**:1941-1944

[13] Boopathy R, Kulpa CF.
Biotransformation of 2,4,6-
trinitrotoluene (TNT) by a *Methano-
coccus* sp. (strain B) isolated from a
lake sediment. Canadian Journal of
Microbiology. 1994;**40**:273-278

[14] Newcombe DA, Crowley
DE. Bioremediation of atrazine-
contaminated soil by repeated
applications of atrazine-degrading
bacteria. Applied and Environmental
Microbiology. 1999;**51**:877-882

[15] Singh BK, Walker A, Morgan
JA, Wright DJ. Biodegradation of
chlorpyrifos by enterobacter strain
B-14 and its use in bioremediation
of contaminated soils. Applied
and Environmental Microbiology.
2004;**70**:4855-4863

[16] Frassinetti S, Setti L, Corti A,
Farrinelli P, Montevecchi P, Vallini
G. Biodegradation of dibenzothiophene
by a nodulating isolate of *Rhizobium
meliloti*. Canadian Journal of
Microbiology. 1998;**44**(3):289-297

[17] Sherburne L, Shrout J, Alvarez P. Hexahydro-1,3,5-trinitro-1,3,5-triazine (RDX) degradation by *Acetobacterium paludosum*. Biodegradation. 2005;**16**:539-547

[18] Zhang C, Hughes JB. Biodegradation pathways of hexahydro-1,3,5-trinitro-1,3,5-triazine (RDX) by clostridium acetobutylicum cell-free extract. Chemosphere. 2003;**50**:665-671

[19] Arun A, Raja P, Arthi R, Ananthi M, Kumar K, Eyini M. Polycyclic aromatic hydrocarbons (pahs) biodegradation by basidiomycetes fungi, pseudomonas isolate, and their cocultures: Comparative in vivo and in silico approach. Applied Biochemistry and Biotechnology. 2008;**151**:2-3

[20] Aitken MD, Stringfellow WT, Nagel RD, Kazunga C, Chen SH. Characteristics of phenanthrene-degrading bacteria isolated from soils contaminated with polycyclic aromatic hydrocarbons. Canadian Journal of Microbiology. 1998;**44**:743-752

[21] Chung SY, Maeda M, Song E, Horikoshi K, Kudo T. A gram-positive polychlorinated biphenyl-degrading bacterium, *Rhodococcus* erythropolis strain TA421, isolated from a termite ecosystem. Bioscience, Biotechnology, and Biochemistry. 1994;**58**:2111-2113

[22] Damaj M, Ahmad D. Biodegradation of polychlorinated biphenyls by rhizobia: A novel finding. Biochemical and Biophysical Research Communications. 1996;**218**:908-915

[23] Atagana HI. Biodegradation of PAHs by fungi in contaminated-soil containing cadmium and nickel ions. African Journal of Biotechnology. 2009;**8**:5780-5789

[24] Dixit R, Wasiullah E, Malaviya D, Pandiyan K, Singh U, Sahu A, et al. Bioremediation of heavy metals from soil and aquatic environment: An overview of principles and criteria of fundamental processes. Sustainability. 2015;**7**(2):2189-2212

[25] Das N, Chandran P. Microbial degradation of petroleum hydrocarbon contaminants: An overview. Biotechnology Research International. 2011:1-13. DOI: 10.4061/2011/941810

[26] Latimer JS, Hoffman EJ, Hoffman G, Fasching JL, Quinn JG. Sources of petroleum hydrocarbons in urban runoff. Water, Air, and Soil Pollution. 1990;**52**:1-21

[27] Husaini A, Roslan HA, Hii KSY, Ang CH. Biodegradation of aliphatic hydrocarbon by indigenous fungi isolated from used motor oil contaminated sites. World Journal of Microbiology and Biotechnology. 2008;**24**:2789-2797

[28] USEPA. Indicators of the Environmental Impacts of Transportation: Highway, Rail, Aviation and Maritime Transport. EPA 230-R-96-009. Washington, DC: US Environmental Protection Agency; 1996. Available from: http://ntl.bts.gov/lib/6000/6300/6333/indicall.pdf

[29] Ward N, Clark J, Lowe P, Seymour S. Water Pollution from Agricultural Pesticides. Centre for Rural Economy Research Report. Newcastle upon Tyne: Centre for Rural Economy; 1993

[30] FWR. What is Pollution? Foundation for Water Research. 2008. Available from: http://www.euwfd.com/html/source_of_pollution_-_overview.html

[31] Van Metre PC, Mahler BJ, Furlong ET. Urban sprawl leaves its PAH signature. Environmental Science & Technology. 2000;**34**:4064-4070

[32] Osuji LC, Nwoye I. An appraisal of the impact of petroleum hydrocarbons on soil fertility: The Owaza experience.

African Journal of Agricultural Research. 2007;2:318-324

[33] Ordinioha B, Brisibe S. The human health implications of crude oil spills in the Niger delta, Nigeria: An interpretation of published studies. Nigerian Medical Journal. 2013;54:10-16

[34] Emmanuel IO, Gordon OD, Nkem AF. The effect of oil spillage on crop yield and farm income in Delta state, Nigeria. Journal of Central European Agriculture. 2006;7:41-48

[35] Abii TA, Nwosu PC. The effect of oil-spillage on the soil of eleme in Rivers state of the Niger-Delta area of Nigeria. Research Journal of Environmental Sciences. 2009;3:316-320

[36] Henry JG, Heinke GW. Environmental Science and Engineering. 2nd ed. New Delhi, India: Prentice Hall; 2005. pp. 64-84

[37] Edema NE, Obadoni BO, Erheni H, Osakwuni UE. Eco-phytochemical studies of plants in a crude oil polluted terrestrial habitat located at Iwhrekan, Ughelli north local government area of Delta state. Natural Science. 2009;7:49-52

[38] Gibson DT, Parales ER. Aromatic hydrocarbon dioxygenases in environmental biotechnology. Current Opinion in Biotechnology. 2000;11:236-243

[39] Urum K, Pekdemi T, Copur M. Surfactants treatment of crude oil contaminated soils. Journal of Colloid and Interface Science. 2004;276:456-464

[40] Mbhele PP. Remediation of Soil and Water Contaminated by Heavy Metals and Hydrocarbons Using Silica Encapsulation. Johannesburg: University of Witwatersrand; 2007

[41] Aguilera F, Mendez J, Pasaro E, Laffon B. Review on the effects of exposure to spilled oils on human health. Journal of Applied Toxicology. 2010;30:291-301

[42] Alonso-Alvarez C, Perez C, Velando A. Effects of acute exposure to heavy fuel oil from the prestige spill on a seabird. Aquatic Toxicology. 2007;84:103-110

[43] Nwilo PC, Badejo OT. Oil spill problems and management in the Niger Delta. International Oil Spill Conference Proceedings. 2005;2005:567-570

[44] Piatt JF, Lensink CJ, Butler W, Kendziorek M, Nysewander DR. Immediate impact of the Exxon Valdez oil spill on marine birds. The Auk. 1990;107:387-397

[45] USFWS. Effects of Oil Spills on Wildlife and Habitat: Alaska Region. U.S. Fish and Wildlife Service. 2004. Available from: http://okaloosa.ifas.ufl.edu/MS/OilSpillFactSheetAlaska.pdf

[46] Balasubramaniam A, Boyle AR, Voulvoulis N. Improving petroleum contaminated land remediation decision making through the MCA weighting process. Chemosphere. 2007;66:791-798. DOI: 10.1016/j.chemosphere.2006.06.039

[47] Nadim F, Hoag GE, Liu S, Carley RJ, Zack P. Detection and remediation of soil aquifer systems contaminated with petroleum products: An overview. Journal of Petroleum Science and Engineering. 2000;26:169-178. DOI: 10.1016/S0920-4105(00)00031-0

[48] Dave D, Ghaly AE. Remediation technologies for marine oil spills: A critical review and comparative analysis. American Journal of Environmental Sciences. 2011;7:423-440

[49] Peer WA, Baxter IR, Richards EL, Freeman JL, Murphy AS. Phytoremediation and hyperaccu-mulator plants. In: Tamas MJ,

Martinoia E, editors. Molecular Biology of Metal Homeostasis and Detoxification. Berlin: Springer; 2006. pp. 299-340. ISBN: 978-3-540-22175-3

[50] Martin W, Nelson YM, Hoffman K. Investigation of hydrocarbon phytoremediation potential of *Lupinus chamissonis* in laboratory microcosms. In: Proceedings of the 77th Annual Water Environment Federation Conference and Exposition; 2-6 October 2004; New Orleans, LA, USA. 2004. pp. 1-26

[51] FRTR. Remediation Technologies Screening Matrix and Reference Guide, Version 4.0. 4.3 Phytoremediation (In situ Soil Remediation Technology). 2012. Available from: http://www.frtr.gov/matrix2/section4/4-3.html

[52] Pivetz BE. Phytoremediation of Contaminated Soil and Ground Water at Hazardous Waste Sites. EPA/540/S-01/500; USEPA. 2001. Available from: http://www.clu-in.org/download/remed/epa_540_s01_500.pdf

[53] Kamath R, Rentz JA, Schnoor JL, Alvarez PJJ. Chapter 16: Phytoremediation of hydrocarbon-contaminated soils: Principles and applications. In: Vazquez-Duhalt R, Quintero-Ramirez R, editors. Petroleum Biotechnology: Developments and Perspectives (Studies in Surface Science and Catalysis). Vol. 151. New York, USA: Elsevier; 2007. pp. 447-478. ISBN-13: 9780080473710

[54] Glick BR. Phytoremediation: Synergistic use of plants and bacteria to clean up the environment. Biotechnology Advances. 2003;**21**:383-393. DOI: 10.1016/S0734-9750(03)00055-7

[55] Huang XD, El-Alawi YS, Penrose D, Glick BR, Greenberg BM. A multiprocess phytoremediation system for removal of polycyclic aromatic hydrocarbons from contaminated soils. Environmental Pollution. 2004;**130**:465-476. DOI: 10.1016/j.envpol.2003.09.031

[56] Huang XD, El-Alawi YS, Gurska J, Glick BR, Greenberg BM. A multiprocess phytoremediation system for decontamination of persistent total petroleum hydrocarbons (TPHs) from soils. Microchemical Journal. 2005;**81**:139-147. DOI: 10.1016/j.microc.2005.01.009

[57] Greenberg BM, Huang XD, Gurska J, Gerhardt KE, Lampi MA, Khalid A, et al. Development and Successful Field Tests of a Multi-Process Phytoremediation System for Decontamination of Persistent Petroleum and Organic Contaminants in Soils. Vol. 1. Canadian Land Reclamation Association; 2006. pp. 124-133

[58] Gerhardt KE, Huang XD, Glick BR, Greenberg BM. Phytoremediation and rhizoremediation of organic soil contaminants: Potential and challenges. Plant Science. 2009;**176**:20-30. DOI: 10.1016/j.plantsci.2008.09.014

[59] Abhilash PC, Jamil S, Singh N. Transgenic plants for enhanced biodegradation and phytoremediation of organic xenobiotics. Biotechnology Advances. 2009;**27**:474-488. DOI: 10.1016/j.biotechadv.2009.04.002

[60] Nzengung VA, Wolfe LN, Rennels DE, McCutcheon SC, Wang C. Use of aquatic plants and algae for decontamination of waters polluted with chlorinated alkanes. International Journal of Phytoremediation. 1999;**1**:203-226. DOI: 10.1080/15226519908500016

[61] Susarla S, Medina VF, McCutcheon SC. Phytoremediation: An ecological solution to organic chemical contamination. Ecological Engineering. 2002;**18**:647-658. DOI: 10.1016/S0925-8574(02)00026-5

[62] Singer AC, Thompson IP, Bailey MJ. The tritrophic trinity: A source of pollutant degrading enzymes and its implication for phytoremediation. Current Opinion in Microbiology. 2004;7:239-244. DOI: 10.1016/j.mib.2004.04.007

[63] Chaudhry Q, Blom-Zandstra M, Gupta SK, Joner E. Utilizing the synergy between plants and rhizosphere microorganisms to enhance breakdown of organic pollutants in the environment. Environmental Science and Pollution Research. 2005;12:34-48. DOI: 10.1065/espr2004.08.213

[64] Marmiroli N, Marmiroli M, Maestri E. Phytoremediation and phytotechnologies: A review for the present and the future. In: Soil and Water Pollution Monitoring, Protection and Remediation. Vol. 69. Dordrecht: Springer; 2006. pp. 403-416. DOI: 10.1007/978-1-4020-4728-2_26

[65] Salanitro JP, Dorn PB, Huesemann MH, Moore KO, Rhodes IA, Ricejackson LM, et al. Crude oil hydrocarbon bioremediation and soil ecotoxicity assessment. Environmental Science and Technology. 1997;31:1769-1776. DOI: 10.1021/es960793i

[66] Thapa B, Kumar KCA, Ghimire A. A review on bioremediation of petroleum hydrocarbon contaminants in soil. Kathmandu University Journal of Science, Engineering and Technology. 2012;8:164-170. DOI: 10.3126/kuset.v8i1.6056

[67] Sharma S. Bioremediation: Features, strategies and applications. Asian Journal of Pharmacy and Life Science. 2012;2:202-213

[68] Doong RA, Wu SC. Substrate effects on the enhanced biotransformation of polychlorinated hydrocarbons under anaerobic condition. Chemosphere. 1995;30:1499-1511. DOI: 10.1016/0045-6535(95)00044-9

[69] Ghazali MF, Rahman RNZA, Salleh AB, Basri M. Biodegradation of hydrocarbons in soil by microbial consortium. International Biodeterioration and Biodegradation. 2004;54:61-67. DOI: 10.1016/j.ibiod.2004.02.002

[70] Medina-Bellver JI, Marin P, Delgado A, Rodríguez-Sánchez A, Reyes E, Ramos JL, et al. Evidence for in situ crude oil biodegradation after the prestige oil spill. Environmental Microbiology. 2005;7:773-779. DOI: 10.1111/j.1462-2920.2005.00742.x

[71] Jones JG, Knight M. Effect of gross population by kerosene hydrocarbons on the microflora of a moorland soil. Nature. 1970;227:1166

[72] Pinholt Y, Struwe S, Kjøller A. Microbial changes during oil decomposition in soil. Holarctic Ecology. 1979;2:195-200. DOI: 10.1111/j.1600-0587.1979.tb00701.x

[73] Mulkins-Phillips GJ, Stewart JE. Distribution of hydrocarbon utilizing bacteria in Northwestern Atlantic waters and coastal sediments. Canadian Journal of Microbiology. 1974:955-962. DOI: 10.1139/m74-147

[74] Patel V, Shah K. Petroleum hydrocarbon pollution and its biodegradation. International Journal of Chemtech Applications. 2014;2:63-80

[75] Gentry TJ, Rensing C, Pepper IL. New approaches for bioaugmentation as a remediation technology. Critical Reviews in Environmental Science and Technology. 2004;34:447-494

[76] Swannell RP, Lee K, McDonagh M. Field evaluations of marine oil spill bioremediation. Microbiological Reviews. 1996;60:342-365

[77] Boufadel MC, Suidan MT, Venosa AD. Tracer studies in laboratory beach

simulating tidal influences. Journal of Environmental Engineering. 2006;**132**:616-623

[78] Zahed MA, Aziz HA, Isa HM, Mohajeri L. Effect of initial oil concentration and dispersant on crude oil biodegradation in contaminated seawater. Bulletin of Environmental Contamination and Toxicology. 2010;**84**:438-442

[79] Lee SH, Lee S, Kim DY, Kim JG. Degradation characteristics of waste lubricants under different nutrient conditions. Journal of Hazardous Materials. 2007;**143**:65-72

[80] Vergetis E. Oil Pollution in Greek Seas and Spill Confrontation Means-Methods. Greece: National Technical University of Athens; 2002

[81] Lessard RR, DeMarco G. The significance of oil spill dispersants. Spill Science and Technology Bulletin. 2000;**6**:59-68

[82] Nomack M, Cleveland CJ. Oil Spill Control Technologies. The Encyclopedia of Earth; 2010. Available from: http://www.eoearth.org/view/article/158385/

[83] USEPA. National Contingency Plan Product Schedule. Washington, DC: US Environmental Protection Agency; 2011 Available from: http://ocean.floridamarine.org/acp/SJACP/Documents/EPA/NCP_Product_Schedule_July_2011.pdf

[84] Fingas MF, Kyle DA, Larouche N, Fieldhouse B, Sergy G, Stoodley G. Effectiveness testing of oil spill-treating agents. In: Lane P, editor. The Use of Chemicals in Oil Spill Response. USA: ASTM International; 1995. pp. 286-298. ISBN: 9780803119994

[85] Watts R, Udell M, Rauch P, Leung S. Treatment of pentachlorophenol-contaminated soils using Fenton's reagent. Hazardous Waste & Hazardous Materials. 1990;**7**:335-345

[86] Karpenko O, Lubenets V, Karpenko E, Novikov V. Chemical oxidants for remediation of contaminated soil and water. A review. Chemistry & Chemical Technology. 2008;**3**:41-45

Chapter 3

Aerosol Studies over Central India

Kannemadugu Hareef Baba Shaeb

Abstract

Earth's radiation budget and thus climate change are significantly influenced by natural and anthropogenic aerosols. Variability of aerosols both in space and time poses challenges to quantify their effects on cloud microphysical properties, precipitation and hydrological cycle. Black carbon (BC) aerosol besides having effects on human health, possess light absorbing nature and thus contribute in atmospheric radiative properties and interaction with clouds. Aerosol properties have been studied over Nagpur (79.028°E, 21.125°N) located in central India, using multi instruments such as multi wavelength radiometer, aethalometer, sunphotometer, balloon based GPS radiosonde, etc., during the study period of 2008–2014. Seasonal variability of different parameters such as aerosol optical depth, columnar water vapor, black carbon mass concentrations, mixed layer height, etc. will be discussed. MODIS aerosol and water vapor products have also been validated against ground based sunphotometer measurements. To understand the source apportionment HYSPLIT model back trajectories have been used. The chapter discusses the interesting aspect of seasonal variability of aerosol properties including monsoonal effects over the data sparse region of central India.

Keywords: aerosol, black carbon, aethalometer, radiosonde, columnar water vapor

1. Introduction

Atmospheric aerosols are solid or liquid particles suspended in air. These particles include sea salt particles, mineral dust, smoke, pollen, etc. Sources of aerosols can be from natural and anthropogenic sources. Generation of aerosols involves individual or combination of chemical, physical and biological processes. Removing processes of aerosols can be of two types, namely wet and dry deposition [1] while dominating process of aerosol removal is wet deposition (cloud and rainfall).

The environment impact assessment of these aerosols is decided by their physical and chemical properties and lifetime. The abundance, size distribution and composition of aerosol particles are highly variable both in space and time. In the lower troposphere, the total particle number concentration (typical mass concentration) typically varies in the range of about 100–100,000 cm^{-3} (1 and 100 $\mu g\ m^{-3}$). In the free troposphere, aerosol concentrations are ($\sim$1–2 orders) of magnitude lower than in the boundary layer.

The aerosol particles can be classified based on the size as the nucleation mode, ultrafine mode and coarse mode. Nucleation mode refers to aerosol particles below 0.1 μm in diameter, whereas diameter is lower than 0.01 μm are called ultrafine mode. The coarse mode refers to particles with diameter larger than 1.0 μm. These particles can accumulate in the atmosphere with lifetime, ranging from 1 to 7 days (boundary layer), 3–10 days (free troposphere) and 1–365 days (in the strato-sphere) and during this period they can undergo long range transport [2].

The chemical composition of atmospheric aerosol consists of variable concentrations of sulphate, nitrate, ammonium, sea salt, crustal elements and carbonaceous compounds (elemental and organic carbon) and other organic materials. Nucleation mode consists of sulphate, nitrate, ammonium, elemental and organic carbon and certain trace metals (e.g., lead, cadmium, nickel, copper, etc.). The coarse mode consists of dust, crustal elements, nitrate, sodium, chloride and biogenic organic particles (e.g., pollen, spores, plant fragments, etc.).

Atmospheric aerosol particles can absorb and scatter the incoming/outgoing shortwave and longwave radiation, which alters the radiation budget of the Earth. The also play important role in the formation of clouds and precipitation since they operate as cloud condensation and ice nuclei. Aerosols can affect significantly the cycles of nitrogen, sulphur, and atmospheric oxidants. Aerosol particles in the upper atmosphere can modify the ozone removal [3]. Additionally aerosols in the lower troposphere affect human health and mortality rate.

The effects of aerosols on climate are very uncertain. Aerosols influence the climate (forcing) in two ways, i.e., direct and indirect. In a direct effect, aerosol particles (especially sulphates) reflect incoming shortwave radiation thus cooling the Earth's atmosphere. However, this cooling effect is compensated by the absorption of longwave terreastrial radiation by absorbing aerosols (black carbon and dust particles). The annual mean radiative forcing (global) is estimated as -0.4 ± 0.2 W m^{-2} (for sulphate), -0.05 ± 0.05 W m^{-2} (for fossil fuel organic carbon), $+0.2 \pm 0.15$ W m^{-2} (for fossil fuel black carbon), $+0.03 \pm 0.12$ W m^{-2} (for biomass burning), -0.1 ± 0.1 W m^{-2} (for nitrate) and -0.1 ± 0.2 W m^{-2} (for mineral dust) [4].

Indirect effect of aerosols affect formation of cloud droplets which are formed by condensation of water vapour onto aerosol particles (cloud condensation nuclei, or ice nuclei) when the relative humidity exceeds the saturation. A very large supersaturation (about 400%) is required for the homogeneous condensation of water vapor in the absnce of aerosols. The increased number of aerosols (i.e., the increased cloud optical thickness) decreases the net surface radiation as they reflect more solar radiation (Twomey effect). Smaller particles can increase cloud lifetime. The absorption of solar radiation by absorbing aerosols can lead to evaporation of cloud particles (semi-direct effect). Anthropogenic aerosols effects on water clouds through the cloud albedo effect cause a negative radiative forcing of -0.3 to -1.8 W m^{-2} [4].

Variability of aerosol parameters over Indian region has been studied using multi-wavelength radiometer (MWR) since 1980 under the Indian Space Research Organization (ISRO) Geosphere Biosphere Program [5]. Aerosol measurements were reported from several places within the country, but such data and results are sparse in a dry tropical region in the central India.

Ground-based observations are important in order to evaluate the accuracy and validity of parameters retrieved from satellites. The validation excerise is usually targeted to the test the retrieval algorithm efficiency and how it can be improved further. India has a wide variety of ecosystems and surface conditions. Hence it is important to validate the satellite-based retrievals using ground-based measurements for different climatic regions throughout the country. Hareef Baba Shaeb et al. [6] reported the validation of the MODIS aerosol optical depth and water vapor over Nagpur located in the central Indian region. Aerosol loading at the measurement site is influenced both by local sources and long range transport. To locate the possible sources, back trajectory analysis is used. We also used MODIS detected fire locations to understand the contribution of biomass burning. They have studied for the first time over this region focusing on the classification of aerosol types, validation of MODIS AOD and water vapor products and the role of aerosol transport. Black carbon (BC) is a primary aerosol emitted directly at the source from incomplete combustion processes such as fossil fuel and biomass burning and therefore much atmospheric BC is of anthropogenic origin (IPCC 2007). BC is receiving much attention recently owing to its effects on weather, atmospheric circulation, and hydrological cycles [7–11] and due to the adverse health impacts of BC [12, 13]. BC possess strong absorption characteristics over wide wavelength range (from UV to near IR) and its chemically inert nature (i.e., longer life time) make this species very important in global change and climate studies [14, 15].

Boundary layer dynamics play important role in surface concentrations of observed BC and its vertical dispersion (convection). The altitude up to which the surface would influence the vertical dispersion of species through convective turbulent eddies is known as the mixed layer height (MLH) and this is an important boundary layer parameter. In view of this Kompalli et al. [16] studied continuous observations of surface BC mass concentration (MBC) along with year-around vertical profiles of atmospheric thermodynamics using balloon borne GPS aided radiosonde ascents from a semi-arid suburban location Nagpur, in Central India are carried out.

The chapter discusses the interesting aspect of seasonal variability of aerosol properties over the data sparse region of central India.

2. Study location and general meteorology

Central India is surrounded by the Great Indian Desert in the northwest, Indo Gangetic Plain in the north and coastal India in east and west. The Nagpur city (21°06′N, 79°03′E; 310 m a.m.s.l) lies at the geographic center of India (**Figure 1**) on the Deccan plateau of Indian peninsula. A very dry and semi humid climate prevails throughout the year except in the monsoon season (June–September).

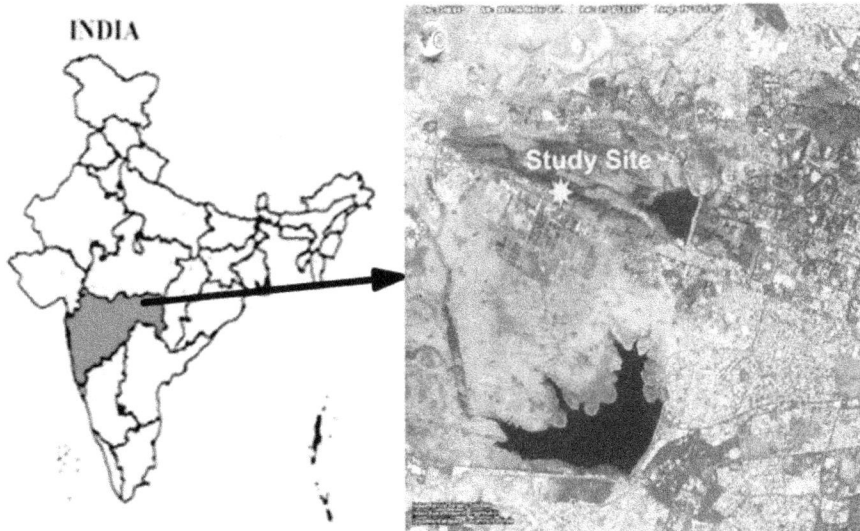

Figure 1.
Study site location.

Dry and hot weather prevails throughout the pre monsoon (PMS) season (March–May). The maximum temperature shoots up to 42–48°C. Summer monsoon (SMS) starts in June and continue up to September. Maximum rainfall is observed during July and August months. During the post monsoon (PoMS) season (October–November), the maximum temperature is about 33°C. Winter season (December, January and February) registers minimum temperatures around 12°C and at times goes below that level.

The monthly mean surface meteorological data, obtained from www.wunderg round.com and rainfall data obtained from www.hydro.imd.gov.in are used for correlating measured Black carbon mass concentrations. **Figure 2(a)** shows

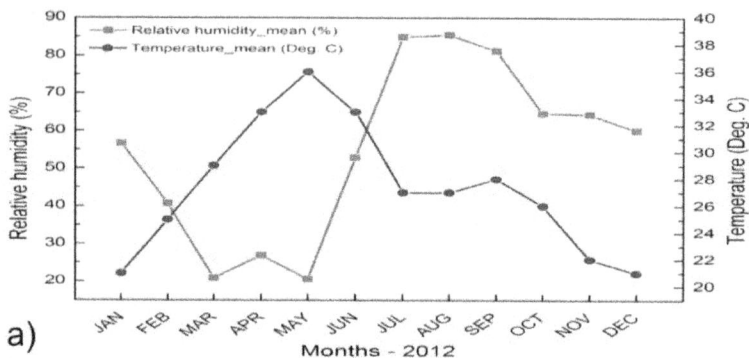

a)

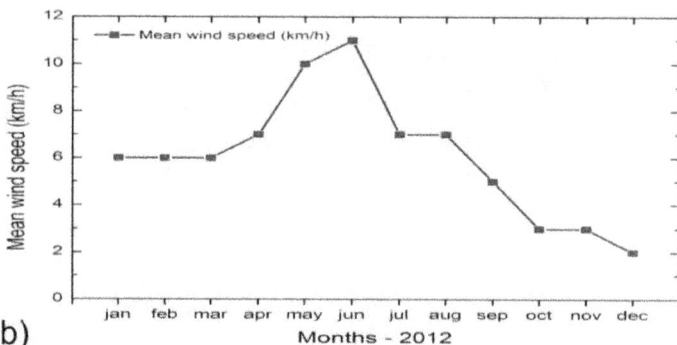

b)

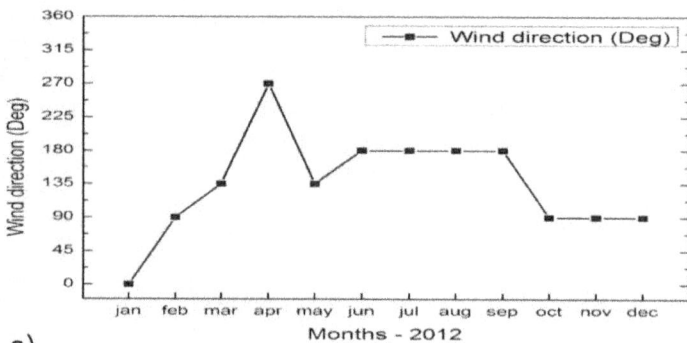

c)

Figure 2.
Monthly variations of (a) relative humidity (%) and temperature (°C), (b) wind speed (m/s) and (c) wind direction (Deg) for 2012.

monthly means of relative humidity and temperature for the year 2012 at the site. Relative humidity is in between ~50–85% during Monsoon and ~60–80% during oost-monsoon and its highest compared to other seasons. Mean wind speed is high during monsoon season and it was observed maximum in the month of June and July. It is gradually increasing from lower values in winter months (not much variation within the winter months) to higher values in pre-monsoon and reaches maximum in monsoon months as high as (2.39 ± 0.53) m/s in the year 2012 decreases in post-monsoon months as like in the winter months. The monthly variation of mean wind speed is shown in **Figure 2(b)** for the years 2012.

Figure 2(c) shows the monthly variation of the mode of wind direction. These wind direction data is used to correlate with cluster trajectories which were used for source appointment. It was observed that mode (maximum number of times) of wind direction is constant about 3–4 months. Thus the wind direction plays a key role in the seasonal transportation of black carbon to study site.

3. Data and methods

3.1 Measurement methods

3.1.1 Ground based

3.1.1.1 Multiwavelength radiometer

Multi-Wavelength Radiometer (MWR) is a passive instrument used for studying the spectral variation in aerosol properties in the visible and near infrared region. The MWR, was mounted on the building upperside, was used to estimate spectral AOD's, on days when unobstructed solar visibility was available for 3 h or a lot of. Aerosol columnar optical depth is calcuable at 10 slender wavelength bands targeted at 380, 400, 450, 500, 600, 650, 750, 850, 935 and 1025 nm. The MWR collects the incoming solar flux as a function of solar zenith angles. The well-known Lambert–Beer–Bouguer Law (Eq. (1)) permits the estimation of AOD, because the output voltage V_λ of the MWR at any wavelength is directly proportional to I_λ, by solving a linear square fit between the logarithm of V_λ and therefore the corresponding relative air mass.

$$I\lambda = I0\lambda \; \exp.[-\tau_\lambda \; mr] \tag{1}$$

where $I_{0\lambda}$ = extra-atmospheric solar irradiance, m_r = relative air mass, $I\lambda$ is the direct solar irradiance at the earth's surface at wavelength λ, τ_λ = total optical thickness. The measured data was edited and further AOD values were calculable following the Langley technique [17, 18]. The total optical depth τ_λ was calculable because the the slope of the curve following the Langley plot methodology. Considering τ_λ as the total of the contribution of the various atmospheric components,

$$\tau_\lambda = \tau_{R\lambda} + \tau_{g\lambda} + \tau_{w\lambda} + \tau_{a\lambda} \tag{2}$$

$\tau_{R\lambda}$ = Rayleigh optical thickness, $\tau_{g\lambda}$ = absorption optical depth (atmospheric gases), $\tau_{w\lambda}$ = optical depth (water vapor), $\tau_{a\lambda}$ = aerosol optical depth. The calculable values of aerosol optical depth $\tau_{a\lambda}$ has errors. The error in τ_λ arises due to 1-min time resolution and the statistical errors in regression calculations. The error in Ozone (O_3) model superimposed with the seasonal differences in O_3 contributes associatean uncertainty of 10% in $\tau_{g\lambda}$ while error in $\tau_{R\lambda}$ is 0.03%. Thus $\tau_{a\lambda}$, may thus have a most application of this methos to the MWR data analysis is described in many earlier papers [5, 19, 20].

The columnar water vapor content has been estimated from the MWR measurements at 935 and 1025 nm [21–23]. The absorption of radiation at 935 nm band is higher by more than three orders of magnitude than at 850 and 1025 nm bands. The details of application of this technique are described by Nair and Moorthy [24].

3.1.1.2 Sun photometer

Model 540 MICROTOPS-II (microprocessor-based Total Ozone Portable Spectrometer) sun photometer is a compact, portable and multi-channel sun photometer is employed to study the characteristics of columnar aerosols properties and columnar water vapor and to validate the satellite retrievals.

The physical and operational characteristics of the instrument are represented within the user's guide (http://www.solar.com/manuals.htm). The sun photometer measures solar irradiance in 5 spectral wave bands (with peak wavelengths of 440, 500, 675, 870, and 936 nm) from that it derives AOD through internal software. The filters utilized in all channels have a peak wavelength preciseness of ±1.5 nm and FWHM band pass of 10 nm (http://www.solar.com/sunphoto.html).

Derivation of AOD and water vapor employing a sun photometer has been clearly explained by Refs. [25, 26]. However; here transient outline is given.

At 440, 500, 675 and 870 nm wavelengths, AOT is derived based on the Beer–Lambert–Bouguer law as follows:

$$V_\lambda = V_{0\lambda}D - 2\exp(-\tau_\lambda M), \tag{3}$$

where, for each channel (wavelength (λ)), V_λ = the signal measured by the instrument, $V_{0\lambda}$ = the extraterrestrial signal, D = Earth-Sun distance in astronomical units, τ_λ = total optical thickness ($\tau_\lambda = \tau_{a\lambda} + \tau_{R\lambda} + \tau_{O3\lambda}$), $\tau_{a\lambda}$ = aerosol optical thickness (AOT), $\tau_{R\lambda}$ = Rayleigh (air) optical thickness, $\tau_{O3\lambda}$ = Ozone optical thickness, M = the optical air mass.

The Rayleigh ($\tau_{R\lambda}$), ozone optical thickness ($\tau O3\lambda$) are obtained from atmospheric models as below:

$$\tau_{R\lambda} = R4 \, \exp.(-h/29.3/273) \tag{4}$$

$$\tau_{O3\lambda} = Ozabs \times DOBS/1000 \tag{5}$$

where h = altitude of the place of observation in meters, R4 = 28773.6 × (R2 × (2 + R2) × λ − 2)2, R2 = 10−8 × {8342.13 + 2,406,030/ (130 − λ − 2) + 15,997/(38.9 − λ − 2)}, λ = wavelength in μm, Ozabs = ozone absorption cross section (extracted from a lookup table based on wavelength), DOBS = ozone amount in Dobson units (extracted from a lookup table based on latitude and date of observation).

MICROTOPS II sun photometer was calibrated by its manufacturer (M/s Solar Light Control, USA) at the Mauna Loa Observatory, Hawaii which is a noise-free high-altitude site before the measurements started in 2011 at our measurement site. Aside from this, we analyzed the MICROTOPS-II output when air mass is eual to zero, which is used as calibration constant. Filter degradation, temperature effects and poor pointing towards the sun can contribute to other measurement errors. The Microtops AOT retrievals uncertainties are in the range of 0.01–0.02 [27].

AERONET stands for Aerosol RObotic NETwork formed by NASA/GSFC and is expanded by collaborators in order to cover a large spatial extent. The sun photometer measurements were performed in cloud-free conditions. For the current study, sun photometer observations are chosen from the condition that the time difference between ground based observation and MODIS flypast time is a smaller amount than quarter-hour. The data set was used because the ground truth within the validation of the Terra Moderate Resolution Imaging Spectroradiometer (MODIS) AOD_{550}.

3.1.1.3 Aethalometer

Aethalometer measures blackcarbon (absorbingaerosol) content by measuring the attenuation of a beam of light transmitted through the sample when collected on a fibrous filter (Lambert–Beer law) at 7 channels (370, 470, 520, 590, 660, 880 and 950 nm). Sixth channel (entered at 880 nm) is considered as the standard channel for BC measurements because BC is the principal absorber of light at this wavelength and other Aerosol components have negligible absorption. Aethalometer (ModelAE-42, Magee Scientific, USA) was operated daily on a 24 h cycle at a flow rate 3 L/min at sampling rate of 5 min interval and air inlet is ~12 m above the ground.

The details of principle of operation, data deduction, error budget of aethalometer, inherent uncertainties in its technique and the corrections are extensively available in the literature (e.g., [28–30]) and are not repeated. The instrumental uncertainty of the aethalometer ranges from 50% at 0.05 μg m^{-3} to 6% at 1 μg m^{-3} [30]. The inherent uncertainties in the aethalometer technique basically arise due to multiple scattering (known as C-factor) and shadowing (R-factor) effects in the filter tape [28–30].

3.1.2 Satellite data

3.1.2.1 MODIS

The MODIS flies on board the EOS Terra and Aqua satellite and measure AOD and other optical properties on a world scale daily from the year 2000 onwards. Terra and Aqua satellites are at an altitude of 705 km, cross equator at 10:30 Indian Standard Time (IST) ascending towards north and at 13:30, IST dropping towards south, respectively. MODIS has 36 bands starting from 0.4 to 14.4 μm wavelengths with three completely different spatial resolutions (250, 500 and 1000 m).

MODIS daily level-3 collection version 005 AOD data at 550 nm averaged at a 1° latitude/longitude grid to produce daily MOD08_D3.005 products from Terra satellite were used. For general climate modeling, the level 3 data provide a convenient source of data that has land and ocean measurements at a 1-degree scale combined into one file. Remer et al. [62] provided international validation of Collection 004 (C004) product over both land and ocean (compared to AERONET) and reported the expected error bars of AOD values as τpλ = ± 0.05 ± 0.15τpλ over land, where τpλ is the AOD value retrieved from the intensity measured at ground. The updated C005 algorithm rule has to be valid, to account for native biases. The aerosol properties contained among the lookup table (LUT) has to be updated for as many ground measuring sites as possible, to improve the accuracy of the retrieved AOD [31].

3.1.2.2 OMI

The Ozone Monitoring Instrument (OMI) is a space-borne spectrometer, which has global coverage on a daily basis with a spatial resolution of 13 × 24 km at nadir. This instrument measures reflected and backscattered solar radiation in UV-visible spectrum (from 250 to 500 nm). Absorbing aerosol index (AAI) or simply aerosol index (AI) is obtained from OMI (http://www.temis.nl/airpollution/absaai/absaai-omi.php?year=2012&datatype=data&freq=daily) gridded daily global level 3 data (NetCDF data format) which is available on ESA Tropospheric Emission Monitoring Internet Service (TEMIS).

The AI is expressed in the following equation:

$$AI = -100 \log \left\{ \left(\frac{I_{\lambda 1}}{I_{\lambda 2}} \right)_{meas} \right\} + 100 \log \left\{ \left[\frac{I_{\lambda 1} \left(A_{LER\lambda 1} \right)}{I_{\lambda 2} \left(A_{LER\lambda 2} \right)} \right]_{calc} \right\} \tag{6}$$

A_{LER} is the surface Lambert equivalent albedo which is dependent on wavelength. AI at 388 nm is obtained using λ_1 (342.5 nm) and λ_2 (388 nm) and is the residue between the measured and calculated radiance assuming Lambert equivalent reflectivity [32, 33]. Tthe presence of absorbing aerosols such as dust and smoke result in positive AI values (>0.2) and high negative values (<−0.2) represent fine non absorbing particles such as sulfates, while AAI values close to zero (±0.2) correspond to clouds or coarse mode non absorbing aerosols [34].

The magnitude of AI is influenced by parameters such as solar zenith angle, aerosol layer height, cloud reflectivity, and pressure but uncertainty/variability can be minimized through seasonal/annual averages [32]. Kascoutis et al. [32] observed that the exclusion of negative AI values may not lead to a true representation of the AI levels at a particular site.

4. Results and discussion

4.1 Seasonal variability in aerosol optical depth

Hareef Baba Shaeb et al. [6] observed that AOD values are observed to be lowest throughout the monsoon because of stronger upper winds, cloud scavenging process and rain wash out [35, 36]. Throughout the post monsoon, aerosols build up slowly and presumably undergo hygroscopic (absorptive) growth in water vapor (RH > 50%) resulting in increase in AOD. In the winter season, AOD exhibits a lot of variability, at first decreasing for the month of December and so steady increasing throughout January and February months. This will be attributed to substantial increase in CWC and temperature from December to January and February. AOD rises in its manitude from winter to summer. High temperature, in association with robust surface winds throughout summer plays a very important role in heating and lifting the top soil layer. This high convective activity and frequent prevalence of long range transport of dust from northwestern India cause increase in AOD throughout this season [37].

Figure 3 shows annual average Moderate Resolution Imaging Instrument (MODIS) Terra AOD_{550} over the Indian subcontinent and surrounding regions. AOD is found to be significant (AOD > 0.7) over northwest, IGP, North east and other parts of India shows relatively less AOD (<0.45).

High AOD_{500} (0.64 ± 0.08) is observed throughout PMS. High temperature, in association with sturdy surface winds, throughout summer plays a vital role in heating and lifting the loose soil. The incursion of wet air either from the Bay of

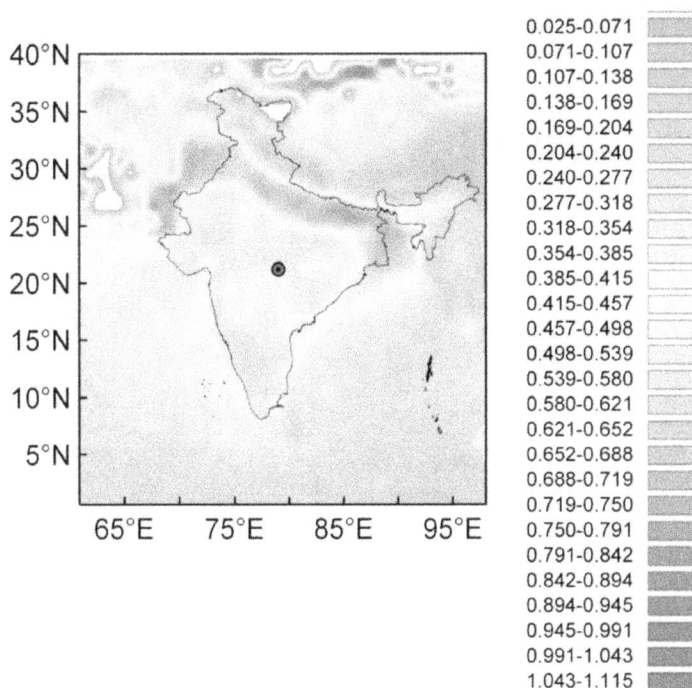

Figure 3.
Annual average moderate resolution imaging instrument (MODIS) Terra AOD_{550} over the Indian subcontinent and surrounding regions.

Bengal or Arabian Sea and/or operation of any trigger mechanism produce conditions contributing for the explosive convective development. This high convective activity and frequent incidence of long range transport of dust from northwestern result in increase in AOD throughout this season [36].

The monsoon typically advances over central India throughout the top of second week or within the third week of June and this can be characterized by severe weather activity i.e., heavy rain, thunderstorm etc. AOD_{500} values (0.38 ± 0.06) are determined to be lower throughout the monsoon season because of stronger higher winds, cloud removal and rain out processes [38]. The withdrawal of monsoon is characterized by the reversal of winds from South West to North East. During the post monsoon, aerosols build up slowly and possibly undergo hygroscopic growth in water vapor (RH > 50%) leading to increase in AOD_{500} (0.5 ± 0.02). The winter season is characterized by dry and cold weather. In the winter season, AOD_{500} is less (0.42 ± 0.15) compared to post monsoon season.

4.2 Seasonal variability in columnar water vapor

A temporal variation of columnar water vapor content (CWC) values for the period from July 2008 to June 2009 is reported by Hareef Baba Shaeb et al. [6]. Minimum columnar water vapor content value of 0.61 g/cm^2 and maximum value of 3.26 g/cm^2 is observed in the months of March 2009 and July 2008 respectively. There exists a well defined seasonal variation in CWC, with the maximum value during the monsoon months and minimum during winter months. Similar variations in columnar water vapor have been observed at other Indian locations [39–42].

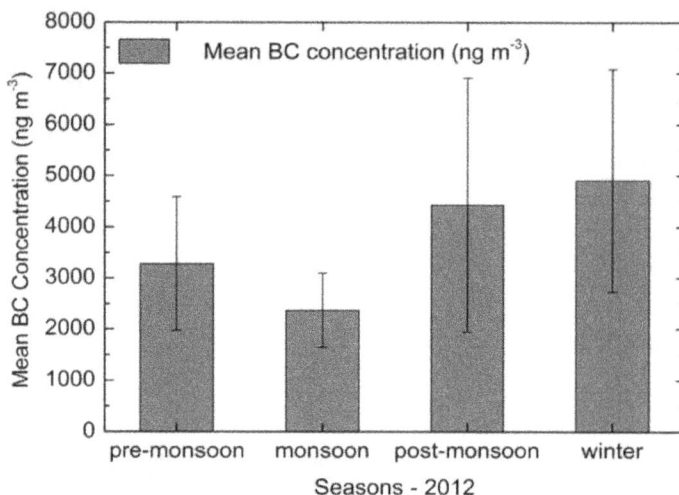

Figure 4.
Seasonal variation of black carbon concentration measured using an Aethalometer at Nagpur, during the year 2012.

It is seen that minimum CWC happens wihin the month of March and starts increasing till July and remains high throughout southwest monsoon months (June–September). CWC starts decreasing shortly once the monsoon season ends from the month of October and then a secondary minimum happens throughout the month of December.

Very good correlation $R^2 \sim 0.7$ ($R^2 \sim 0.5$) is observed between RH (%) and CWC in summer and the monsoon (post monsoon and winter). This signifies the correlation between near surface and columnar water vapor amounts.

4.3 Seasonal variations in BC mass concentration

Seasonal variations of BC aerosol mass concentration showed high values, during the post monsoon (4.4 ± 0.9 µg m^{-3}) followed by winter (4.2 ± 0.6 µg m^{-3}) season and low values during the monsoon (2.4 ± 0.6 µg m^{-3}) followed by pre monsoon (3.3 ± 0.6 µg m^{-3}) season. The variation is shown in the form of bar chart in **Figure 4**.

The annual average BC concentration is found to be 3.57 ± 0.7 µg m^{-3} and this is 20% more than the value found for the year 2011. High values of wind speed (and total rain fall) during monsoon and pre monsoon seasons may be responsible for observed low values of BC mass concentrations. During winter and post monsoon low temperatures (which keep mixed layer height low), low relative humidity leads to observed high concentrations at the surface level.

4.4 Comparisons with other locations in India

The BC mass concentrations have been compared with the measurements reported from other locations in India. This value (3.57 ± 0.7 µg m^{-3}) is lower compared to urban areas like Ahmedabad, Pune and much lower in comparison to urban and industrial locations like Delhi and Mumbai (**Table 1**).

Station	Location/ environment	Period	Mean M_{BC} ($\mu g\ m^{-3}$)	Reference
Srinagar (34.06 °N, 74.78°E)	Northern India/urban	Jan 2013-Dec 2013	6	[43]
Darjeeling (27.03° N, 88.26°E)	Eastern India	Jan 2010 to Dec 2011	3.45	[44]
Dayalbagh, Agra (27.23°N, 78.0026° E)	Northern India (Indo-Gangetic basin)/urban	May 2014 to April 2015	9.5	[45]
Kanpur (26.46°N, 80.32°E)	Northern India (central part of IGP)/ urban	8 January 2015 to 28 February 2015	4.06 ± 2.46	[46]
Ooty (11.4°N, 76.7° E)	South India (Western ghats)/	April 2010 to May 2012	0.96 ± 0.35 (summer) 0.23 ± 0.06 (monsoon)	[47]
Ahmedabad (23.03°N, 72.55°E)	Western India/urban	Winter	11.6 ± 2.9	[48]
Mumbai (19.13°N, 72.91°E)	Western Coast/urban industrial	January to March 1999	12.5	[49]
Ananthapur (14.36°N, 77.65°E)	Southern plateau/rural (semi-arid)	August 2006 to July 2007	1.97	[50]
Hyderabad (17.47° N, 78.58°E)	South-Central India/ urban	January to July 2003	0.5–68 (dry season) 0.5–45 (wet season)	[51]
Kharagpur (22.5°N, 87.3°E)	Eastern coast of North India/industrialized	2004–2008	6.50 ± 3.04	[52]
New Delhi (28.63° N, 77.17°E)	North India/urban industrial Southern	May 2001 to April 2002	17.9 (6.7–27.9)	[53]
Trivandrum (8.5° N, 77°E)	Sothern peninsular semiurban/coastal	August 2000 to October 2001	0.3–5	[54]
Pune (18.53°N, 73.85°E)	Western/urban	January to December 2005	4.1	[55]

Table 1.
BC values reported by different authors.

4.5 Seasonal variations in mixed layer depth

In order to estimate mixed layer height (MLH), the raw data on temperature, pressure, relative humidity and geographical position (latitude, longitude and altitude) as a function of time at every 1 s, are filtered and regridded at 10 m regular interval. The top of mixed layer is defined as the altitude where the vertical gradients inθv exceeded 3 K km^{-1}. The equations used and procedure to obtain MLH is explained in Ref. [16] hence not repeated. The mean mixed layer height values for PMS, SMS, PoMS and winter are found to be 3014 ± 1187, 832 ± 452, 1871 ± 506 and 1488 ± 706 m respectively, therfore showing the least values in monsoon season, highest values in pre monsoon and moderate to low values in post monsoon and winter seasons [16].

According to Ref. [16] the main conclusion from association of MLH with BC is a good association between MLH and MBC was seen during dry period of the year (winter and PMS). However, during wet period the association between $M_{BC}$and MLH is low.

4.6 Seasonal variability in absorbing aerosol index

Annual mean variation of AAI for the year 2011 is shown in **Figure 5**. The positive values (>0.2) which represent absorbing aerosols such as dust is present in north western region (Thar desert region) and it extends even to IGP region though with less concentrations. Over Southern India the AAI values are negative indicating lesser influence of dust related aerosol particles.

Monthly mean variation of AAI at Nagpur is shown in **Figure 6**. AAI values are highly positive (>0.2) during pre-monsoon months (Mar, April, May) indicating dominance of absorbing aerosols such as dust while highly negative (<−0.2) during Monsoon (Jun, Jul, Aug, Sept) indicating the presence of non-absorbing aerosols such as sulfates. During winter (Dec, Jan, Feb) the AAI values close to zero (±0.2) indicates the presence of clouds or coarse mode non absorbing aerosols.

4.7 Hysplit back trajectories

Hareef et al. [56] analyzed the airmass back trajectories in association with forest fires for various seasons specifically, PMS, SMS, PoMS and winter. Analysis urged that in PMS, the air masses were started from the biomass burning regions, desert

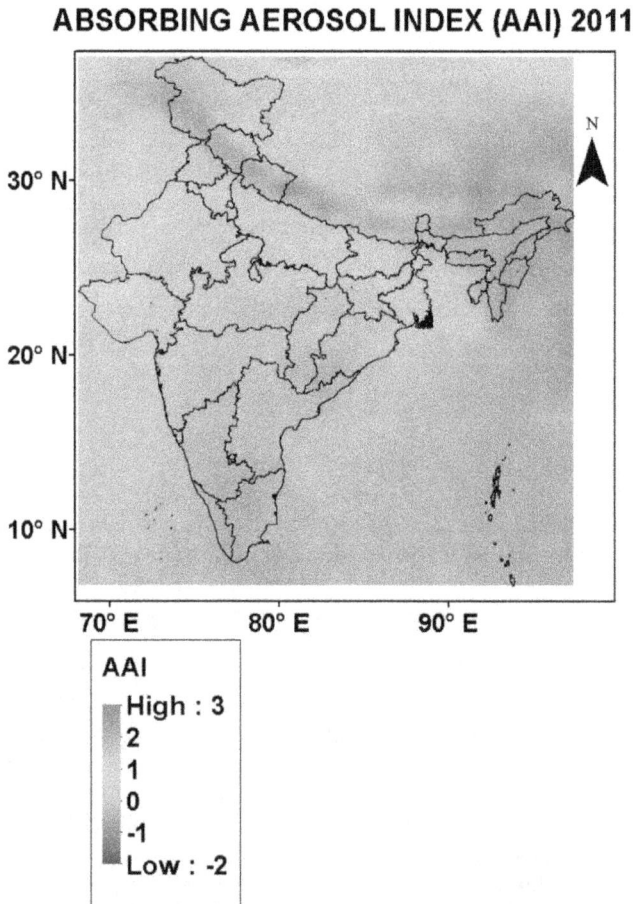

Figure 5.
Annual mean variation of AAI over India.

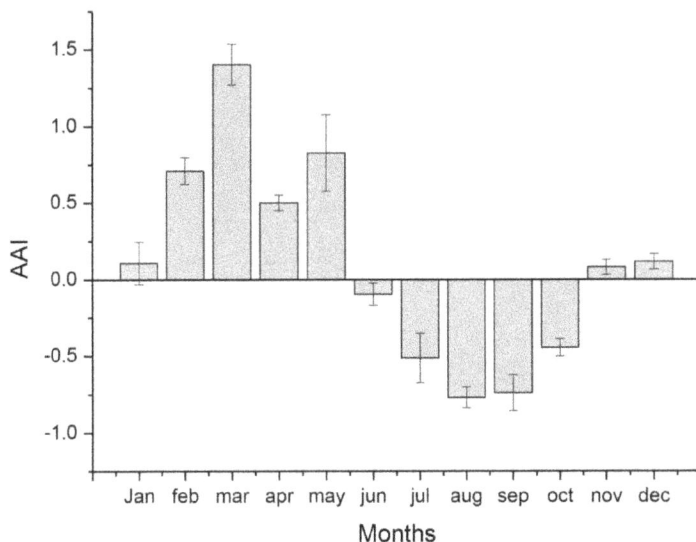

Figure 6.
Monthly variation of AAI at Nagpur.

regions and also from marine regions. During SMS, as a result of the sustained south westerly flow of the monsoon winds, the air masses were largely of marine origin. Throughout the post monsoon season, the dominant air masses were started from north India, as well as transport of air masses from biomass burning, i.e., Punjab region. Throughout winter, the origins of air masses were set in eastern India and IGP region. **Figure 7** shows example back trajectory starting at 09:00 UTC on 12 Aug 2012 at the Nagpur location.

4.8 Comparison with MODIS aerosol and water vapor products

The geophysical parameters retrieved from satellites need to be validated against the ground measurements in order to understand the retrieval errors and to correct them accordingly. This validation exercise needs to be performed for different surfaces globally. Towards this, detailed validation of MODIS AOD products of different versions with distinct spatial resolutions by using the ground-based multi wavelength radiometer and MICROTOPS sun photometer has been performed by several authors over the Indian subcontinent (e.g., [57–61]). The studies found MODIS overestimating the AOD values during the summer and underestimating during winter. Hareef et al. [56] validated the MODIS aerosol product version C005 over the central Indian region where there is no validation exercise done so far. Authors found a high correlation of 0.75 observed indicates that the MODIS can capture the seasonal variability well, and a slope of 0.65 implies an underestimation of 35% lower AOD compared to sun photometer. In the MODIS AOD retrieval algorithm, by default neutral aerosol model (Single Scattering Albedo (SSA) ~ 0.9) was set for a major part of Asia [62, 63] for different seasons in a year. Absorbing (SSA ~ 0.85) or non-absorbing (SSA ~ 0.95) models were applied in rest of the world. This is supported by the aerosol varieties determined in AERONET sites situated at different parts around the world and supported the condition that If either the non-absorbing or the absorbing aerosol occupied more than 40% of the pie, and the other occupied less than 20%, then the location was selected as the dominant aerosol type. In India, there is just one AERONET site (IIT Kanpur: 26.28° N, 80.24°E) situated inside the IGP region and aerosol varieties determined there is

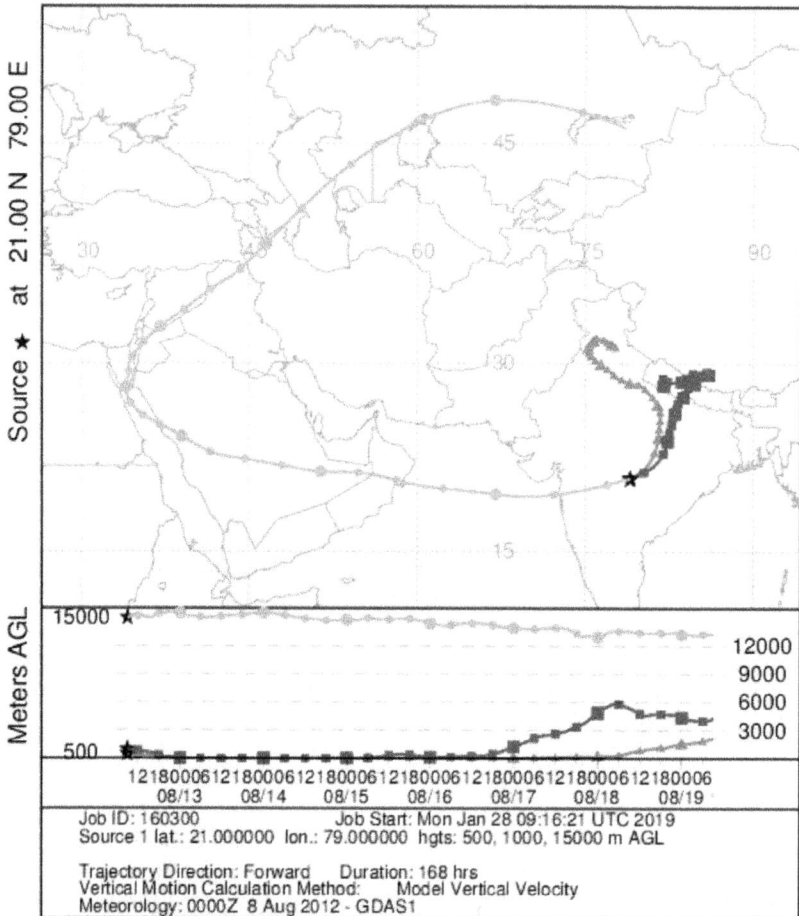

Figure 7.
Forward trajectories starting at 09:00 UTC on 12 Aug 2012 at the Nagpur location.

used in the retrieval of AOD for other locations. During PMS main aerosol types observed over Nagpur location, were UB and DD, however MODIS algorithm assumes neutral aerosol whereas there is a good proportion of UB is present that is of absorbing type. The absolute error between AOD measured (the sun photometer) to that of MODIS retrieved AOD is maximum (0.29) for this season. A decent variety of MODIS fire locations over central India and back trajectories additionally indicate that there is a transport from such places. This can cause underestimation of AOD, as for absorbing aerosols if the algorithm assumes the scattering aerosols, it will incorrectly assign a smaller AOD value to match calculated radiance with determined radiance resulting in underestimation of the AOD (retrieved from satellite). The Single Scattered Albedo for black carbon aerosols because of biomass burning is considerably not up to that of dust particles. This might be the rationale, whereas different authors reported overestimation of MODIS AOD compared to ground measured AOD, over this region, Hareef et al. [56] had observed the underestimation because of the significant amount of black carbon.

The authors concluded that the MODIS aerosol optical depth retrievals do not represent, accurately, true observations in central India, and therefore cannot be well applied there. This could be attributed to a complex nature of surface conditions and aerosol varieties and seasonal nature of surface reflectance and aerosol models over completely different ecological and geographic regions. Thus we recommend better absorbing type of model and conjointly embody seasonally dynamic changing land use/land cover options in central India for correct retrieval.

Hareef et al. [56] reported columnar water vapor (CWC) amount measured using a sun photometer over this region as typically in the 0.4–4 cm range. There exists a well-defined seasonal variation in CWC, with the maximum value during the monsoon months and minimum during winter months. Similar variations in CWC have been observed from other locations in India [5, 40]. The validation of gridded products (MODIS) is important as they are used for assimilation in numerical weather prediction and global climate models [64]. For this purpose, detailed validation of MODIS water vapor product is attempted. Validation of MODIS TERRA retrieved water vapor (NIR) with Sun photometer suggests 20% overestimation by MODIS with correlation coefficient 0.89, which has been attributed to errors due to turbidity or haze in the atmosphere.

5. Conclusions

This chapter presents the aerosol studies over Nagpur, a tropical station in central India. The main conclusions of the study are summarized as follows:

1. AOD showed highest value (0.64 ± 0.08) during the summer, while lowest during the monsoon season (0.38 ± 0.06).

2. There exists a well-defined seasonal variation in columnar water vapor content (CWC), with the maximum value during the monsoon months and minimum during winter months. Columnar water vapor (CWC) amount measured using a sun photometer over this region as typically in the 0.4–4 cm range.

3. Comparison of AOD (MODIS) and water vapor (NIR) (MODIS), with the sun photometer observations, indicates an underestimation of 35% lower AOD (correlation coefficient $\sim$0.75) and overestimation of 20% higher water vapor (correlation coefficient $\sim$0.89) respectively.

4. Aerosol transport analysis suggests during PMS, the air masses were originated from the biomass burning regions, desert regions and also from marine regions.

5. Seasonal variations of BC aerosol mass concentration showed high values, during the post monsoon ($4.4 \pm 0.9 \ \mu g \ m^{-3}$) followed by winter ($4.2 \pm 0.6 \ \mu g \ m^{-3}$) season and low values during the monsoon ($2.4 \pm 0.6 \ \mu g \ m^{-3}$) followed by pre monsoon ($3.3 \pm 0.6 \ \mu g \ m^{-3}$) season.

6. The BC mass concentrations have been compared with the measurements reported from other locations in India indicating the lower value compared to urban areas like Ahmedabad, Pune and much lower in comparison to urban and industrial locations like Delhi and Mumbai.

7. Seasonally the mean MLH values show the lowest in monsoon, highest values in PMS and moderate to low values in PoMS and winter.

8. AAI values are highly positive during pre-monsoon months, indicating dominance of absorbing aerosols such as dust while highly negative during monsoon indicating the presence of non-absorbing aerosols such as sulfates. During winter the AAI values close to zero indicates the presence of clouds or coarse mode non absorbing aerosols.

Acknowledgements

Most of the work presented in this chapter is carried out as a part of Aerosol Radiative Forcing Over India (ARFI) project of ISRO-GBP, author thanks Space Physics Laboratory for conceiving and executing such a wonderful project. Author thank Dr. Dibyendu Datta, Group Director, Climate Sciences Group, Dr. Sesha Sai M.V.R, Deputy Director, Earth and Climate Science Area and Director, NRSC for the support and encouragement towards this study. I am grateful to the NOAA Air Resources Laboratory (ARL) for the provision of the HYSPLIT transport and dispersion model through their website https://ready.arl.noaa.gov/HYSPLIT.

Conflict of interest

The author declares that there is no conflict of interest.

Author details

Kannemadugu Hareef Baba Shaeb
Earth and Climate Science Area, ISRO-Department of Space, National Remote Sensing Centre, Hyderabad, Telangana, India

*Address all correspondence to: babaphyiway@gmail.com

IntechOpen

References

[1] Petroff A, Mailliat A, Amielh M, Anselmet F. Aerosol dry deposition on vegetative canopies. Part I: Review of present knowledge. Atmospheric Environment. 2008;**42**:3625-3653

[2] Jaenicke R. Atmospheric aerosols and global climate. Journal of Aerosol Sciences. 1980;**11**:577-588

[3] Mészáros E. Fundamentals of Atmospheric Aerosol Chemistry. Budapest: Akadémiai Kiadó; 2000. p. 308. ISBN: 9630576246

[4] Solomon S, Qin D, Manning M, Chen Z, Marquis M, Averyt KB, et al., editors. Climate Change 2007: The Physical Science Basis. Contribution of Working Group I to the Fourth Assessment Report of the Intergovernmental Panel on Climate Change. Cambridge, United Kingdom and New York, NY, USA: Cambridge University Press. p. 996. ISBN: 978 0521 88009-1

[5] Moorthy KK, Niranjan K, Narasimhamurthy B, Agashe VV, Murthy BVK. Aerosol climatology over India. In: 1-ISRO GBP MWR Network and Database, Scientific Report, ISRO GBP, SR 03. Vol. 99. India: Indian Space Research Organization; 1999

[6] Hareef Baba Shaeb K, Joshi AK, Moharil SV. Aerosol optical properties and types over Nagpur, Central India. Sustainable Environment Research. 2014;**24**(1):29-40

[7] Menon S, Hansen J, Nazarenko L, Luo Y. Climate effects of blackcarbon aerosols in China and India. Science. 2002;**297**:2250-2253. DOI: 10.1126/science.1075159

[8] Lau KM, Kim MK, Kim KM. Asian summer monsoon anomalies induced by aerosol direct forcing: The role of the Tibetan Plateau. Climate Dynamics. 2006;**26**:855-864. DOI: 10.1007/s00382-006-0114-z

[9] Meehl GA, Arblaster JM, Collins WD. Effects of black carbon aerosols on the Indian monsoon. Journal of Climate. 2008;**21**:2869-2882

[10] Gautam R, Hsu NC, Lau KM, Kafatos M. Aerosol and rainfall variability over the Indian monsoon region: Distributions, trends and coupling. Annales de Geophysique. 2009;**27**:3691-3703

[11] Lawrence MG, Lelieveld J. Atmospheric pollutant outflow from southern Asia: A review. Atmospheric Chemistry and Physics. 2010;**10**:11017-11096. DOI: 10.5194/acp-10-11017-2010

[12] Janssen NA, Hoek G, Simic-Lawson M, Fischer P, van Bree L, ten Brink H, et al. Black carbon as an additional indicator of the adverse health effects of airborne particles compared with PM10 and PM 2.5. Environmental Health Perspectives. 2011;**119**:1691-1699. DOI: 10.1289/ehp.1003369

[13] Janssen NAH, Gerlofs-Nijland ME, Lanki T, Salonen RO, Cassee F, Hoek G, et al. Health Effects of Black Carbon. Denmark: World Health Organization; 2012. ISBN: 978 92 890 0265 3

[14] Bond TC, Doherty SJ, Fahey DW, Forster PM, Berntsen T, DeAngelo BJ, et al. Bounding the role of black carbon in the climate system: A scientific assessment. Journal of Geophysical Research. 2013;**118**. DOI: 10.1002/jgrd.50171

[15] Zha S, Cheng T, Tao J, Zhang R, Chen J, Zhang Y, et al. Characteristics and relevant remote sources of black carbon aerosol in Shanghai. Atmospheric Research. 2014;**135–136**:159-171

[16] Kompalli SK, Suresh Babu S, Krishna Moorthy K, Manoj MR, Kiran Kumar NVP, Hareef Baba Shaeb K, et al. Aerosol black carbon characteristics over Central India: Temporal variation and its dependence on mixed layer height. Atmospheric Research. 2014; **147-148**(523):27-37

[17] Shaw GE, Regan JA, Herman BM. Investigations of atmospheric extinctions using direct solar radiation measurements made with a multiple wavelength radiometer. Journal of Applied Meteorology. 1973:12374-12380

[18] Moorthy KK, Nair PR, Krishna Murthy BV. Multi wavelength solar radiometer network and features of aerosol spectral optical depth at Trivandrum. Indian Journal of Radio and Space Physics. 1989;**18**:194-120

[19] Moorthy KK, Saha A, Prasad BSN, Niranjan K, Jhurry D, Pillai PS. Aerosol optical depths over peninsular India and adjoining oceans during the INDOEX campaigns: Spatial, temporal and spectral characteristics. Journal of Geophysical Research. 2001;**106**: 28539-28554

[20] Saha A, Moorthy KK. Impact of precipitation on aerosol spectral optical depth and retrieved size distributions: A case study. Journal of Applied Meteorology. 2004;**43**(6):902-914

[21] Leckner B. The spectral distribution of solar radiation at the earth's surface— Elements of a model. Solar Energy. 1978: 20143-20150

[22] Kneizys FX, Shettle EP, Gallery WO, Chetwynd JH Jr, Abreu LW, Selby JEA, et al. Atmospheric Transmittance/ Radiance: Computer Code, Lowtran 5. AFGL-TR-80-0067. MA, USA: AFGL; 1980

[23] Tanaka M, Nakazawa T, Fukabori M. Absorption of the ρστ, 0.8 um and α bands of the water vapour. Journal of

Quantitative Spectroscopy and Radiative Transfer. 1982;**28**:463-470

[24] Nair PR, Moorthy KK. Effects of changes in the atmospheric water vapour content on the physical properties of atmospheric aerosols at a coastal station. Journal of Atmospheric and Solar Terrestrial Physics. 1998;**60**: 563-572

[25] Ichoku C, Levy R, Kaufman YJ, Remer LA, Li RR, Martins VJ, et al. Analysis of the performance characteristics of the five-channel microtops II sun photometer for measuring aerosol optical thickness and precipitable water vapor. Journal of Geophysical Research. 2001;**106**: 14573-14582

[26] Morys M, Mims FM III, Hagerup S, Anderson SE, Baker A, Kia J, et al. Design, calibration, and performance of microtops II handheld ozone monitor and sun photometer. Journal of Geophysical Research. 2001;**106**: 14573-14582

[27] Holben BN, Tanre D, Smirnov A, Eck TF, Slutsker I, Abuhassan N, et al. An emerging ground-based aerosol climatology: Aerosol optical depth from AERONET. Journal of Geophysical Research. 2001;**106**:12067-12097

[28] Weingartner E, Saathoff H, Schnaiter M, Streit N, Bitnar B, Baltensperger U. Absorption of light by soot particles: Determination of the absorption coefficient by means of aethalometers. Journal of Aerosol Science. 2003;**34**:1445-1463

[29] Arnott WP, Hamasha K, Moosmuller H, Sheridan PJ, Ogren JA. Towards aerosol light-absorption measurements with a 7-wavelength aethalometer: Evaluation with a photoacoustic instrument and 3-wavelength nephelometer. Aerospace Science and Technology. 2005;**39**(1): 17-29

[30] Corrigan CE, Ramanathan V, Schauer JJ. Impact of monsoon transition on the physical and optical properties of aerosols. Journal of Geophysical Research. 2006;**111**: D18208

[31] Levy RC, Remer LA, Kleidman RG, Mattoo S, Ichoku C, Kahn R, et al. Global evaluation of the collection 5 MODIS dark-target aerosol products over land. Atmospheric Chemistry and Physics. 2010;**10**:10399-10420

[32] Kaskaoutis DG, Nastos PT, Kosmopoulos PG, Kambezidis HD. The combined use of satellite data, air-mass trajectories and model applications for monitoring of the dust transport over Athens Greece. International Journal of Remote Sensing. 2010;**31**:5089-5109. DOI: 10.1080/01431160903283868.

[33] Herman JR, Bhartia PK, Torres O, Hsu C, Seftor C, Celarier E. Global distribution of UV-absorbing aerosols from Nimbus 7/TOMS data. Journal of Geophysical Research. 1997;**102**: 16911-16922

[34] Torres O, Tanskanen A, Veihelmann B, Ahn C, Braak R, Bhartia PK, et al. Aerosols and surface UV products from ozone monitoring instrument observations: An overview. Journal of Geophysical Research. 2007; **112**:D24S47. DOI: 10.1029/ 2007JD008809

[35] Kaskaoutis DG, Badarinath KVS, Kharol SK, Sharma AR, Kambezidis HD. Variations in the aerosol optical properties and types over the tropical urban site of Hyderabad, India. Journal of Geophysical Research. 2009;**114**: D22204. DOI: 10.1029/2009JD012423

[36] Flossman FI, Hall WD, Pruppacher HR. A theoretical study of the wet removal of atmospheric pollutants— Part I: The redistribution of aerosol particles captured through nucleation and impaction scavenging by growing

cloud drops. Journal of the Atmospheric Sciences. 1985;**42**:583-606

[37] Badarinath KVS, Kharol SK, Kaskaoutis DG, Kambezidis HD. Dust storm over Indian region and its impact on the ground reaching solar radiation case study using multi-satellite data and ground measurements. Science of the Total Environment. 2007;**384**: 316-332

[38] Balakrishnaiah G, Raghavendra Kumar K, Suresh Kumar Reddy B, Rama Gopal K, Reddy RR, Reddy LSS, et al. Analysis of optical properties of atmospheric aerosols inferred from spectral AODs and Angstrom wavelength exponent. Atmospheric Environment. 2011;**45**:1275-1285

[39] Ernest Raj P, Devara PCS, Saha SK, Sonbawne SM, Dani KK, Pandithurai G. Temporal variations in sun photometer measured precipitable water in near IR band and its comparison with model estimates at a tropical Indian station. Atmosfera. 2008;**21**(4):317-333

[40] Ranjan RR, Ganguly ND, Joshi HP, Iyer KN. Study of aerosol optical depth and precipitable water vapour content at Rajkot, a tropical semi arid station. Indian Journal of Radio and Space Physics. 2007;**36**:27-32

[41] IMAP FINAL REPORT-III. Characteristics of Aerosol Spectral Optical Depths over India, ISRO-IMP-SR-43-94. 1994. pp. 1-78

[42] Bhat MA, Romshoo SA, Beig G. Aerosol black carbon at an urban site-Srinagar, Northwestern Himalaya, India: Seasonality, sources, meteorology and radiative forcing. Atmospheric Environment. 2017;**165**:336-348

[43] Sarkar C, Chatterjee A, Singh AK, Ghosh SK, Raha S. Characterization of black carbon aerosols over Darjeeling— A high altitude Himalayan station in Eastern India. Aerosol and Air Quality

Research. 2015;**15**:465-478. DOI:
10.4209/aaqr.2014.02.0028

[44] Gupta P, Singh SP, Jangid A, Kumar
R. Characterization of black carbon in
the ambient air of Agra, India: Seasonal
variation and meteorological influence.
Advances in Atmospheric Sciences.
2017;**34**(9):1082-1094. Available from:
https://link.springer.com/article/
10.1007/s00376-017-6234-z

[45] Navaneeth M, Thamban SNT,
Moosakutty SP, Kuntamukkala P,
Kanawade VP. Internally mixed black
carbon in the Indo-Gangetic Plain and
its effect on absorption enhancement.
Atmospheric Research. 2017;**197**:
211-223. Available from: http://home.
iitk.ac.in/~snt/pdf/Thamban_AR_2017.
pdf

[46] Udayasoorian C, Jayabalakrishnan
RM, Suguna AR, Gogoi MM, Suresh
Babu S. Aerosol black carbon
characteristics over a high-altitude
Western Ghats location in Southern
India. Annales de Geophysique. 2014;**32**:
1361-1371. DOI: 10.5194/angeo-32-1361-
2014

[47] Ramachandran S, Kedia S. Black
carbon aerosols over an urban region:
Radiative forcing and climate impact.
Journal of Geophysical Research. 2010;
115:D10202. DOI: 10.1029/
2009JD013560. Available from: https://
agupubs.onlinelibrary.wiley.com/doi/
pdf/10.1029/2009JD013560

[48] Venkatraman C, Habib G, Eiguren-
Fernandez A, Mignel AH, Friedlander
SK. Residential biofuels in South Asia:
Carbonaceous aerosol emissions and
climate impacts. Science. 2005;**307**:
1454-1456

[49] Kumar KR, Narasimhulu K,
Balakrishnaiah G, Reddy BSK, Gopal
KR, Reddy RR, et al. Characterization of
aerosol black carbon over a tropical
semi-arid region of Anantapur, India.

Atmospheric Research. 2011;**12–27**
(530):100

[50] Latha KM, Badarinath KVS. Black
carbon aerosols over tropical urban
environment—A case study.
Atmospheric Research. 2003;**69**:125-133

[51] Rai K, Sarkar AK, Mitra AP.
Chemical characterization of aerosols.
IASTA Bulletin. Delhi: NPL; 2002;**14**:
155-158

[52] Kompalli SK, Moorthy KK, Babu SS.
Rapid response of atmospheric BC to
anthropogenic sources: Observational
evidence. Atmospheric Science Letters.
2013;**15**(3):166-171. DOI: 10.1002/
asl2.483. Published online in wileyonline
library.com

[53] Rai K, Sarkar AK, Mitra AP.
Chemical characterization of aerosols at
NPL, Delhi. Proceedings of a conference
on aerosol remote sensing in global
change and atmospheric pollution.
IASTA Bulletin. 2002, 1982;**14/1**(special
issue):155-158

[54] Babu SS, Moorthy KK. Aerosol black
carbon over a tropical coastal station in
India. Geophysical Research Letters.
2002;**29**(23):131e-1341e

[55] Safai PD, Kewat S, Praveen PS, Rao
PSP, Momin GA, Ali K, et al. Seasonal
variation of black carbon aerosols over a
tropical urban city of Pune, India.
Atmospheric Environment. 2007;
41(13):2699-2709

[56] Hareef Baba Shaeb K, Varghese AO,
Mukkara SR, Joshi AK, Moharil SV.
Aerosol type's classification and
validation of MODIS aerosol and water
vapor products using a sun photometer
over central India. Aerosol and Air
Quality Research (AAQR). 2015;**15**(2):
682-693

[57] Tripathi SN, Dey S, Chandel A,
Srivastava S, Singh RP, Holben BN.
Comparison of MODIS and AERONET

derived aerosol optical depth over the
Ganga Basin, India. Annales de
Geophysique. 2005;**23**:1093-1101

[58] Prasad AK, Singh S, Chauhan SS,
Srivastava MK, Singh RP, Singh R.
Aerosol radiative forcing over the Indo-
Gangetic Plains during major dust
storms. Atmospheric Environment.
2007;**41**:6289-6301

[59] Misra A, Jayaraman A, Ganguly D.
Validation of MODIS derived aerosol
optical depth over Western India.
Journal of Geophysical Research. 2008;
113:D04203

[60] Vinoj V, Satheesh SK, Moorthy KK.
Aerosol characteristics at a remote
Island: Minicoy in Southern Arabian
Sea. Journal of Earth System Science.
2008;**117**:389-397

[61] Guleria RP, Kuniyal JC, Rawat PS,
Thakur HK, Sharma M, Sharma NL,
et al. Validation of MODIS retrieval
aerosol optical depth and an
investigation of aerosol transport over
Mohal in North Western Indian
Himalaya. International Journal of
Remote Sensing. 2012;**33**:5379-5401

[62] Remer L, Kaufman YJ, Tanre D,
Mattoo S, Chu DA, Martins J, et al. The
MODIS aerosol algorithm, products, and
validation. Journal of the Atmospheric
Sciences. 2005;**62**:947-973

[63] Levy RC, Remer LA, Dubovik O.
Global aerosol optical properties and
application to moderate resolution
imaging spectroradiometer aerosol
retrieval over land. Journal of
Geophysical Research. 2007;**112**:1-15

[64] Prasad AK, Singh RP. Validation of
MODIS Terra, AIRS, NCEP/DOE AMIP-
II reanalysis-2, and AERONET Sun
photometer derived integrated
precipitable water vapor using ground-
based GPS receivers over India. Journal
of Geophysical Research. 2009;**114**:
D05107

Chapter 4

Recent Advances for Polycyclic Aromatic Analysis in Airborne Particulate Matter

Hugo Saldarriaga-Noreña, Rebecca López-Márquez,
Mario Alfonso Murillo-Tovar, Mónica Ivonne Arias-Montoya,
Jorge Antonio Guerrero-Álvarez and Josefina Vergara-Sánchez

Abstract

Polycyclic aromatic hydrocarbons (PAHs) are formed in natural processes during combustion of biomass (e.g., forest fires) and by anthropogenic activities at high temperatures. In according with the suggestion the major sources of PAHs in the environment. The main sources of PAHs come basically from heat and power generation (e.g., coal, gas, wood, and oil), industrial processes (e.g., coke production), refuse burning and vehicle emissions. Human exposure to airborne PAHs can result from these processes, as well as from emissions from other sources, such as cooking, smoking, and materials containing PAHs (e.g., petroleum products and fuels). The potential serious health effects resulting from acute and chronic human exposure to PAHs are of concern. For this reason, the identification and quantification of PAHs in airborne particles have been a real challenge, given the multiple impacts that these substances represent for human health. In the last decade, multiple technological developments have been implemented, ranging from sampling systems, extraction and analysis of these compounds with the aim of obtaining more accurate and reliable results. This chapter was prepared to describe and to assess the state of the art about the evolution and application of sampling, extraction and analysis methodologies for the determination of PAHs in airborne particles.

Keywords: airborne particles, PAHs, cleanup, GC-MS/MS

1. Introduction

Polycyclic aromatic hydrocarbons (PAHs) comprise a large variety of organic compounds whose main characteristic is that they are formed by the fusion of benzene rings [1]. PAHs are originated mainly from incomplete pyrolysis of organic materials. Pyrolysis is the process in which organic compounds such as fuels undergo a change in the molecular structure at high temperature without sufficient oxygen concentration. These reactions are mainly dependent on temperature and concentration and are generally endothermic [2].

During combustion at high temperatures and relatively low amounts of oxygen, part of the combustible material is fragmented into small molecular masses, usually to free radicals by pyrolysis (approximately 500–800°C), which recombine to give

53

rise to the PAHs by pyrosynthesis by decreasing the temperature. Once PAH of low molecular weight is formed, (e.g., naphthalene, 128) the pyrosynthesis process continues with "zigzag additions," which generates high molecular weight PAH [3].

PAHs are categorized as low molecular weight (LMW) and high molecular weight (HMW) based on molecular structure. The LMW PAHs include two and three rings structure while HMW PAHs comprise four and more rings structure. The carcinogenicity of PAHs increased with increasing molecular weight [4].

PAHs are ubiquitous pollutants in the atmosphere. The behavior of PAHs in the atmosphere depends on complex physicochemical reactions, interactions with other pollutants, photochemical transformations, and dry and wet deposition. PAHs in the ambient air exist in vapor phase or adsorb into airborne particulate matter depending on the atmospheric conditions (ambient temperature, relative humidity, etc.), the nature (i.e., origin and properties) of the aerosol, and the properties of the individual PAH [5]. The physicochemical properties of PAHs make them highly mobile in the environment, allowing them to distribute across air, soil, and water bodies where their presence is ubiquitous. PAHs are widely distributed in the atmosphere. The PAHs entering the atmosphere can be transported over long distances before deposition through atmospheric precipitation onto soils, vegetation or water [5].

The adsorption of PAHs onto particulate phases can also be affected by the humidity. Moreover, PAH adsorption also depends on the types of suspended particulates (e.g., soot, dust, fly ash, pyrogenic metal oxides, pollens, etc.) and the amounts of dust in the air influence PAH concentrations in the particulate phase [5].

PAHs are known to be toxic and carcinogenic [6]. They are metabolized in the body through oxidation by P450 enzymes and may produce carcinogenic metabolites. These metabolites have been shown to induce lung and skin tumors in animals [7]. People can be exposed through polluted air from urban or industrial environments, tobacco smoke, and diet [6]. The carcinogenicity of the PAHs usually increases with increased number of aromatic rings and higher molecular weight, while low molecular weight PAHs are more acute toxic [7]. Many hundreds of PAHs exist in the environment, but the US Environmental Protection Agency (USEPA) has listed 16 as "Consent Decree" priority pollutants chosen because, because of the likelier risk to be exposed to them, the high amount of information about them, and that they are believed to be more harmful [7] (**Table 1**).

PAH compound (s)	MW (g/mol)	CAS	Molecular formula	Carcinogenic classification	Agency	Structure
Naphthalene	128	91-20-3	$C_{10}H_8$	Not classifiable as to human carcinogenicity	U.S. Environmental Protection Agency (EPA)	
Acenaphthylene	152	208-96-8	$C_{12}H_8$	Not classifiable as to human carcinogenicity	U.S. Environmental Protection Agency (EPA)	
Acenaphthene	154	83-32-9	$C_{12}H_{10}$	It has not been studied yet	U.S. Environmental Protection Agency (EPA)	
Fluorene	166	86-76-7	$C_{13}H_{10}$	Not classifiable as to their carcinogenicity to humans	International Agency for Research on Cancer (IARC)	

PAH compound (s)	MW (g/mol)	CAS	Molecular formula	Carcinogenic classification	Agency	Structure
Phenanthrene	178	85-01-8	$C_{14}H_{10}$	Not classifiable as to their carcinogenicity to humans	International Agency for Research on Cancer (IARC)	
Anthracene	178	120-12-7	$C_{14}H_{10}$	Not classifiable as to their carcinogenicity tohumans	International Agency for Research on Cancer (IARC)	
Fluoranthene	202	206-44-0	$C_{16}H_{10}$	Not classifiable as to their carcinogenicity to humans	International Agency for Research on Cancer (IARC)	
Pyrene	202	129-00-0	$C_{16}H_{10}$	Not classifiable as to their carcinogenicity to humans	International Agency for Research on Cancer (IARC)	
Benzo(a) anthracene	228	56-55-3	$C_{18}H_{12}$	Probably carcinogenic to humans	International Agency for Research on Cancer (IARC)	
Chrysene	228	218-01-9	$C_{18}H_{12}$	Not classifiable as to their carcinogenicity to humans	International Agency for Research on Cancer (IARC)	
Benzo(b) fluoranthene	252	205-99-2	$C_{20}H_{12}$	Probable human carcinogens	U.S. Environmental Protection Agency (EPA)	
Benzo(k) fluoranthene	252	207-08-9	$C_{20}H_{12}$	Probable human carcinogens	U.S. Environmental Protection Agency (EPA)	
Benzo(a)pyrene	252	50-32-8	$C_{20}H_{12}$	Probable human carcinogens	U.S. Environmental Protection Agency (EPA)	
Indeno(1 2 3-cd)pyrene	276	193-39-5	$C_{22}H_{12}$	Probable human carcinogens	U.S. Environmental Protection Agency (EPA)	
Dibenzo(a,h) anthracene	278	53-07-3	$C_{22}H_{14}$	Probable human carcinogens	U.S. Environmental Protection Agency (EPA)	
Benzo(g,h,i) perylene	276	191-24-2	$C_{22}H_{12}$	Not classifiable as to their carcinogenicity to humans	International Agency for Research on Cancer (IARC)	

Table 1.
Some characteristics and carcinogenic classification for PAHs.

2. Methodology and analysis of polycyclic aromatic compounds in airborne particles

2.1 Sampling

The partition of PAHs between gas and particulate phases in the atmosphere fundamentally depends on the vapor pressure, temperature, atmospheric pressure, and the concentration [8, 9]. PAHs having two rings exist in the gas phase, PAHs having three and four rings are in both phases and PAHs having five rings or more exist in the particle phase [10].

The standard methods to measure PAHs in ambient air are active samplers, and these equipment use a pump to draw the air into the sampler, through the filter and the following adsorbent.

The samplers have a sampling module which often consists of two compartments: a filter and a solid adsorbent to collect the particle associated and the gas phase pollutants, respectively. The filter, often teflon, glass or quartz fiber, is placed in the inlet of the sampler [11–15]. The solid adsorbent normally consists of a polyurethane foam (PUF) plug or a sorbent tube with XAD-2 or Tenax depending on the target pollutants and the capacity required; the adsorbent also retains pollutants that volatilize from the particles on the filter during sampling.

The other alternative is passive samplers, in contrast to active samplers, not in need of a pump and electricity to collect pollutants. Instead, the collection is based on a free flow of pollutants from the air to the collecting medium. Most of the existing passive samplers are designed for gas sampling of semi-volatile organic compounds and based on high capacity sampling against a linear sampling rate for long durations such as weeks or months. Polyurethane (SPMDs), XAD-resin based samplers and membrane samplers are such examples [16–18].

In conclusion, sampling equipment can be active or passive, must be compatible and consistent with the analysis method and the monitoring objectives.

2.2 Extraction and cleanup

Given the complexity of environmental matrices, specialized analytical procedures are required for the determination of PAHs. The analytical procedures must include different stages for extraction of compounds from complex samples, purification and detection techniques for multicomponent mixtures that consist of compounds with a wide range of molecular weights, volatilities and polarities.

The extraction of PAHs from airborne particulates is mainly done through methods based on the use of solvents [19]. Being soxleth and ultrasound techniques the most commonly used for extracting soluble organic matter [20]. Subsequently, other solvent-based methods have been developed, accelerated solvent extraction and microwave extraction, both methods have the characteristics that use less solvent and extraction time [21, 22]. Finally, solid phase microextraction (SPME) was adapted for the extraction of PAHs associated with airborne particles, specifically for those of low molecular weight (less than four rings). The main characteristic of this method is that it uses very small amounts of solvents compared to the other extraction techniques mentioned [23]. **Table 2** shows a summary of the main techniques for the extraction of organic compounds from environmental matrices, some characteristics and applications.

PAHs extracts from airborne particulate matter represent a very complex matrix in trace amounts, which contain saturated hydrocarbons, nitrogen, oxygen, and

Techniques	Characteristics	Analytes	References
Soxhlet	It has been so far applied for organic compound extraction from solid matrices due to its high extraction efficiency	PAHs, polybrominated diphenyl ethers (PBDEs), polychlorinated biphenyls (PCBs), among others	[29–31]
Ultrasound assisted extraction (UAE)	Ultrasound energy has also been widely used for the leaching of organic and inorganic compounds from solid matrices	Pharmaceutical Endocrine disruptor compounds (EDCs), perfluorochemicals (PFCs), antibiotics, tetrabromobisphenol-A (TBBPA) PAHs, phthalate esters (PEs), PCBs, nonylphenols (NPs), nonylphenol ethoxylates (NPEOs) and pharmaceuticals and personal care products	[32–35]
Pressurized liquid extraction (PLE)	Pressure is applied to allow the use of extraction solvents or mixtures at temperatures higher than their normal boiling point. The increase on the extraction temperature can promote higher analyte solubility by increasing both solubility and mass transfer rate	Perfluorinated acids (PFAs), perfluorosulfonates (PFSs) and perfluoroctanesulfonamide PAHs	[36, 37]
Microwave-assisted extraction (MAE)	Microwave-assisted extraction (MAE) uses microwave energy to heat the sample-solvent mixture. This technique reduces the extraction times and the extractant amount for the extraction of organic compounds from solid matrices	PAHs in airborne particles, 17-estradiol (E2), estriol (E3), 17-ethinyl estradiol (EE2)	[38, 39]
Supercritical fluid extraction (SFE)	Supercritical fluid extraction is an alternative extraction method with the advantages of reduced solvent consumption and extraction time compared with the classical extraction techniques. Carbon dioxide is commonly used as fluid and methanol is added as organic modifier when polar compounds are extracted	PAHs in marine sediment	[40, 41]
Solid Phase Microextraction (SPME)	It is used specifically for the extraction of low molecular weight organic compounds, from liquid, air and solid matrix	PAHs or polybrominated biphenyls (PBBs)	[42]

Table 2.
Classification of the main extraction techniques, characteristics and applications.

sulfur heterocompounds, among others, what difficult the identification of the PAHs identification in environmental samples [24–26]. After liquid extraction, a cleanup procedure is recommended to eliminate some interferences that can affect the PAHs detection in the chromatographic analysis. The most common cleanup procedures are as follows: liquid-liquid extraction (LLE) and solid phase extraction (SPE) with silica gel and/or C18 cartridges [27, 28].

2.3 Chromatographic analysis

2.3.1 Gas chromatography coupling to mass spectrometer (GC-MS)

Traditionally, the analysis of PAHs in environmental samples has been carried out by gas chromatography (GC), rather than liquid chromatography (LC), and this is due to its greater selectivity, resolution and sensitivity. GC-MS is one of the most powerful analytical tools available for the chemical analysis of complex mixtures. The use of mass spectrometry enhances the capabilities of gas chromatography; the specific information provided in the mass spectrum makes the mass spectrometer a highly selective detector that can be used for qualitative analysis and structural determination.

Mass spectrometry is undoubtedly one of the most widely used for the analysis and characterization of chemical compounds due to its high sensitivity and resolution capacity. Then, when it is coupled to the chromatographic techniques, it is particularly useful in the identification and quantification of organic substances of interest, which are in trace concentrations in environmental samples, it is highly valued mainly for its high sensitivity, that is, it is feasible to quantify those substances contained in a sample quantity of the order of mg.

One of the main disadvantages of mass spectrometry is the conditions used for the generation of stable ionized species and their adequate detection. This process can be a limitation to clearly observe the molecular ion and, therefore, perform a detailed analysis of the chemical structure. In the literature, a significant number of ionization methods have been reported, which depend on the physicochemical properties of the system in question. However, ionization is not the only limitation in mass technique; there are previous problems such as the distinction of different structures when analyzing complex mixtures. Therefore, the use of chromatography as a separation technique is essential for environmental samples such as those obtained in the ambient air, while after the ionization is the analyzer that is where the ionized species are separated and detected, they are also of great importance for analysis.

The criteria considered the most relevant in mass spectrometry are as follows: sensitivity, resolution, stability and selectivity; that depending on the level we need to reach each one of these, there are several types of coupled mass spectrometry equipment arrangements, so their choice depends mainly on the different chemical systems and the scope of the analysis. In the case of air samples such as gases and respirable suspended particles, the unambiguous quantification and identification of the analyte are the main objectives; therefore, the standardization and prior validation of the method used are essential, so it is necessary to carry out several preliminary tests with specific equipment arrangements in order to achieve reliable results.

The mass spectrometer is among the most sensitive chromatographic detectors, having a detection limit below the picogram level, through the use of selected ion monitoring (SIM) mode. PAHs are easily resolved using standard GC columns without a requirement for derivatization. Most separations can be achieved in less than 30 min using capillary columns such as 30 m × 0.25 mm i.d. and 0.25 μm film thickness, 5% phenyl polysiloxane type phases. The use of a narrow bore, thin film column allows an increase in chromatographic resolving power, coupled with a reduction in analysis time.

GC-MS method has been used to all or some subset of the US Environmental Protection Agency (US-EPA) 16 priority PAHs. Single quadrupole GC-MS has offered the opportunity to increase selectivity for these analytes over that of classical detectors, such as UV and fluorescence detectors in high pressure liquid chromatography (HPLC) and electron capture detector (ECD) and flame ionization detector (FID) detectors in GC. This has allowed for limited optimization of sample preparation procedures to increase time to result [43, 44].

Quadrupole mass analyzers are widely used in many areas of environmental analysis. Although popularly referred to as quadrupole mass spectrometers, the mass-resolving properties of such devices are really much more similar to those of a tunable variable hand pass mass filter. Only ions within a narrow mass region (generally <1 amu) are allowed to pass through the device.

Quadrupole mass analyzers have several advantages such as no requirement for very high vacuum (>10^{-7} Torr) and their relatively fast and simple operation for high-throughput analysis. Disadvantages include low transmittance, a low m/z cutoff, and low (generally unit) resolution. Electron impact (EI) is well established and is the most common method of ionization in gas chromatography (GC) [45]. The molecules exiting the gas chromatograph are bombarded by an electron beam (70 eV), which removes an electron from the molecule resulting in a charged ion.

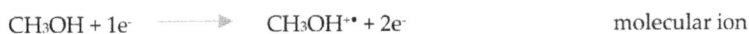

$$CH_3OH + 1e^- \longrightarrow CH_3OH^{+\bullet} + 2e^- \qquad\qquad \text{molecular ion}$$

EI mode produces single charged molecular ions and fragment ions, which are used for structure elucidation.

$$CH_3OH^{+\bullet} \dashrightarrow CH_2OH^+ + H^\bullet \text{ or } CH_3^+ + OH^\bullet \qquad \text{fragment ion}$$

The generated mass spectrum plots the signal intensity at a given m/z ratio (**Figure 1**).

Mass spectrometric methods are particularly suited for analysis of PAHs, because these compounds are semi-volatile and occur as complex mixtures; electron ionization (EI) and chemical ionization (CI) with quadrupole or magnetic sector mass spectrometers have been effectively used to determine PAHs. Although distinguishing between the isomeric forms of PAHs using EI is difficult because the isomers tend to produce common intermediates that give identical losses upon high-energy ionization or collisional activation [46].

For example, when trying to distinguish between compounds with the same or similar molecular weight, the GC-MS coupling is difficult, given its low resolution, **Figures 2** and **3** show the separation of chrysene and triphenylene and mass spectrum respectively. The quantification of chrysene is often biased due to its coelution with triphenylene (the compounds with m/z 228 also contain fragments of m/z 226). Another case is the separation of the isomers of benzo(b, k, j)fluoranthene (252 m/z).

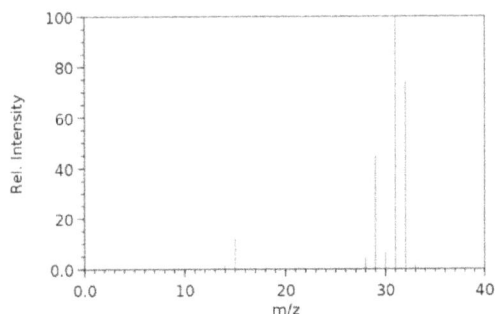

Figure 1.
Mass spectrum for methanol obtained by electronic impact [47].

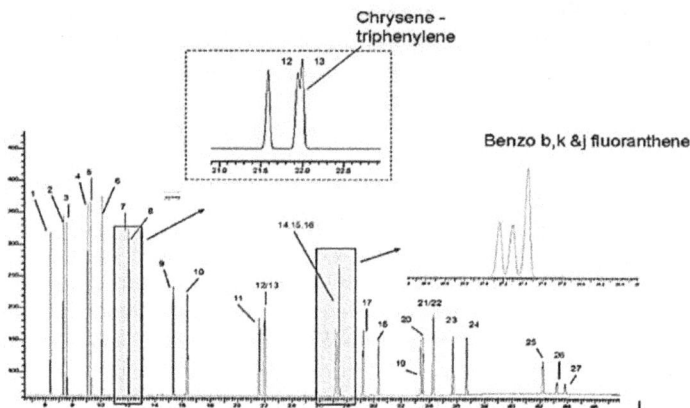

Figure 2.
Examples for PAHs separations with similar molecular weight [47].

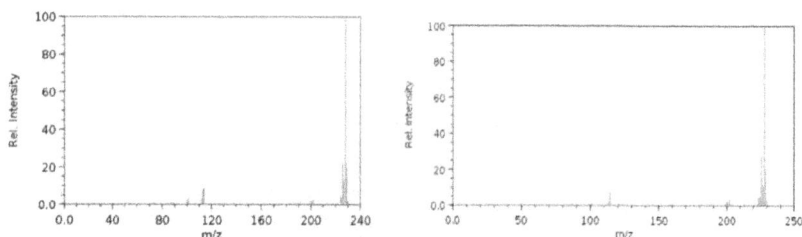

Figure 3.
Mass spectrum for chrysene and triphenylene [47].

2.3.2 Gas chromatography triple quadrupole

As seen in the previous examples, the mass spectrum with a single quadrupole is limited to distinguish among compounds that have structural isomers, since it only considers a single identification criterion, in fact they show a similar mass to charge fragmentation patterns (**Figure 4**). Recently, the coupling gas chromatography coupled to triple quadrupole mass spectrometry (GC-QqQ) was developed. Comparing triple quadrupole analyzer (QqQ) with single quadrupole analyzer, the product ion is more specific than the ion in the simple MS spectrum because the tandem configuration offers the only alternative of selecting the precursor ion of each compound by

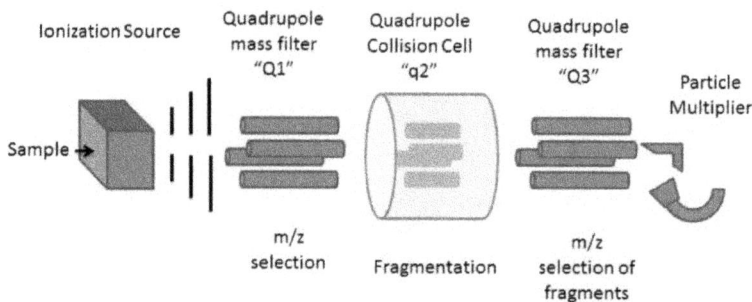

Figure 4.
Triple quadrupole mass analyzer (QqQ) [49].

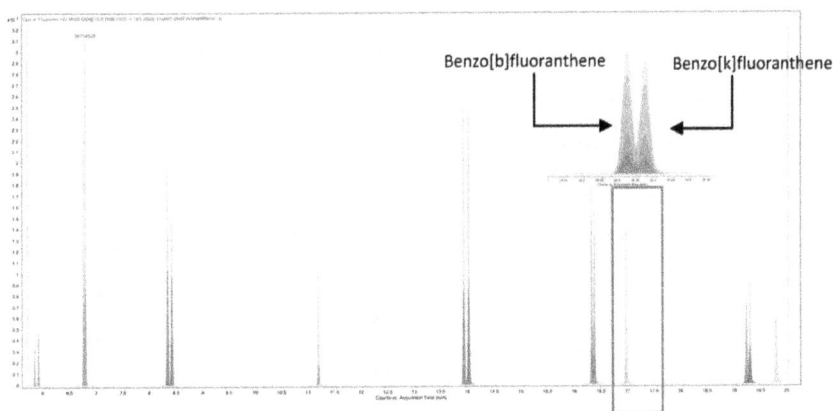

Figure 5.
Chromatogram 14 PAHs by GC-QqQ (own authorship).

the first quadrupole and filtering it into the collision cell, with the consequent elimination of the remaining fragments and consequently the decrease in noise (**Figure 5**). Then that mass to charge pattern obtained by the second spectrometer, which is derived from the collision of the parent fragment, this usually has a unique pattern mass to charge daughter that provides invaluable structural information of the substance, which decreases the probability of false positives and facilitates the unequivocal identification of the target compound. It is clear that the coupling substantially solves the difficulties of simple couplings and can significantly improve the reliability of the determination by offering lower noise levels and additional identification criteria. Several reports show the utility of different arrays of mass spectrometers coupled to gas and liquid chromatographs used for the PAHs and their derivatives analysis. Among the most used is the triple quadrupole GC–MS/MS system, which provides detection and quantification levels equivalent to parts per trillion.

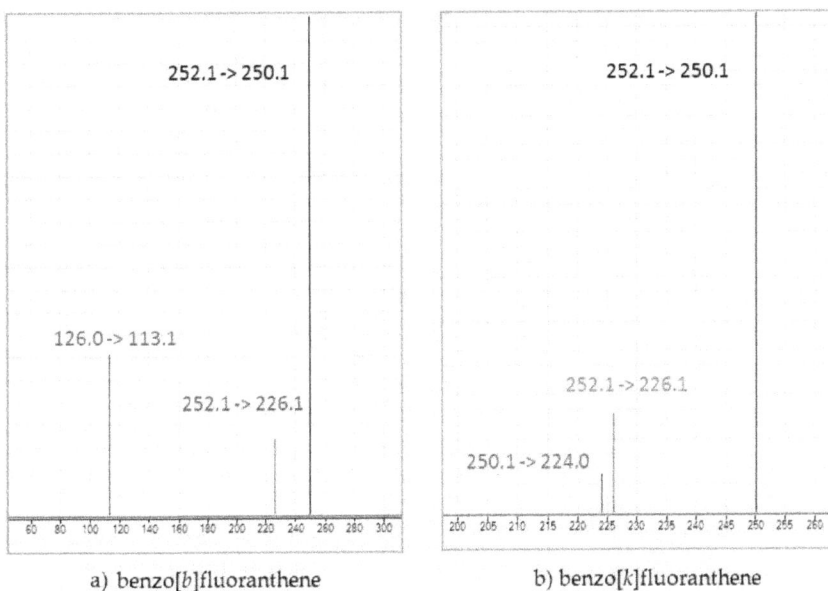

a) benzo[*b*]fluoranthene   b) benzo[*k*]fluoranthene

Figure 6.
Mass to charge (m/z) transitions (a) and (b).

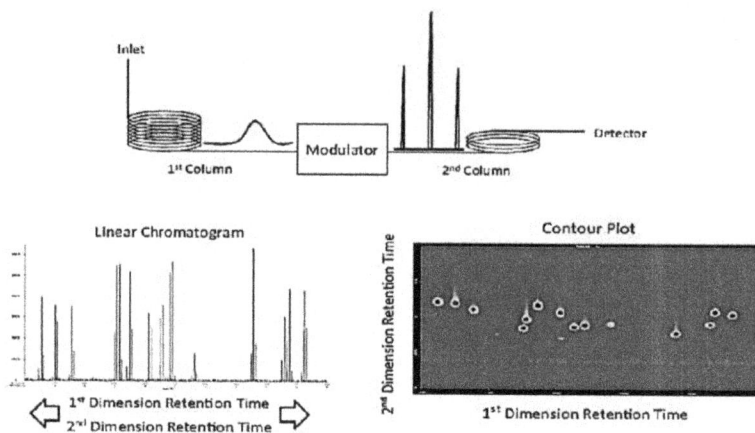

Figure 7.
Scheme of the gas chromatography bidimensional (GC × GC): peaks eluting from the first dimension column enter the second dimension column through the modulator [60].

For instance, the unequivocal identification of B(b)F and B(k)F isomers (**Figure 6**) was successfully achieved by distinct transitions mass to charge obtained with triple quadrupolar mass spectrometer arrange (**Figure 7**). It means QqQ may provide more accurate quantification and confirmation in trace analysis with complex matrix [48].

For these characteristics, recently the GC-QqQ has been proposed for the quantification of the PAHs and its derivatives. Its characteristics have proved to be useful for example in the determination of nitro-PAHs in PM_{10} particles obtained with low sample volume (16.7 lpm). This implies a lower mass of the compound per gram of particle collected per day, and without an exhaustive treatment and sample purification need as those that are followed for particles obtained from high volume samplers (1.3 m^3/h). An analytical method was recently developed for the simultaneous determination of 14 nitro-PAHs (2-nitrofluorene, 9-nitroanthracene, 9-nitrophenanthrene, 3-nitrophenanthrene, 2-nitroanthracene, 3-nitrofluoranthene, 1-nitropyrene, 2,7-dinitrofluorene, 7-nitrobenzo [a]anthracene, 6-nitrochrysene, 1,3-dinitropyrene, 1,8-dinitropyrene, 1,6-dinitropyrene, and 6-nitrobenzo[a] pyrene) in PM_{10} by GC-QqQ in multiple reaction monitoring (MRM) mode. The method performance evaluation showed that the technique is quite reliable, since it provides high repeatability with relative standard deviation <10% and with detection limits between 0.25 and 10 ng/mL. This was also facilitated its application to only half of the filter containing the sample, in this way the remaining part served to complement the chemical characterization of the sample [50]. In the MRM mode, as the name implies, in the first quadrupole (Q1) one or multiple precursor ions of the analyzed substance were are filtered, which react by fragmenting in the collision cell (Q2), until arriving at the second quadrupole (Q3), where the ions product of the quantification and qualification are filtered. The results of this type of analysis are highly specific and sensitive because they provide unique structural information of the molecule that leads to its identification and unambiguous distinction between other substances contained in the sample [50].

Gas chromatography with tandem mass spectrometry has also been successfully applied for the determination of precursor PAHs. Although they are found in environmental levels between one and two orders of magnitude higher than their derivatives, they are trace concentrations substances, which are similarly affected by the different interferences that may come in the samples and by the matrix effect in the extracts. Therefore, the organic extract must be purified before its analysis by GC-MS [51]. However, it was shown that when samples obtained with high volume equipment are analyzed, the extraction is sufficient, and the purification of the extract can be dispensed with before its analysis by GC-MS/MS [52].

Another coupling proposed for the analysis of the nitro-PAHs and oxy-PAHs derivatives is the ultrahigh pressure liquid chromatography-atmospheric pressure chemical ionization-tandem mass spectrometer (UHPLC-(+)-APCI-MS/MS). In addition to the stated advantage of the tandem arrangement, this alternative technique aims to contribute to reducing the thermal degradation that has been consistently reported for those oxy-PAHs classified as quinones in the injection port of the GC. Thus, facilitating their simultaneous analysis with the nitro-PAHs, and taking advantage of the improvements in the sensitivity and selectivity in the determination in organic and aqueous extracts obtained from $PM_{2.5}$ and PM_{10} particles. It has also been found that chemical ionization at atmospheric pressure (APCI), and photoionization at atmospheric pressure provides high ionization efficiency for oxy-PAHs, while electrospray ionization efficiency is usually lower [53]. In a pioneering study, it was shown that liquid chromatography atmospheric pressure chemical ionization-tandem mass spectrometer (LC-APCI-MS/MS) is feasible for the determination of oxy-PAHs and can contribute to the simplification of sample preparation by reducing it to an extraction and evaporation step [54]. Consistently, in a study of the simultaneous analysis of 5 nitro-PAHs—1-nitropyrene (1-NPYR), 2-nitrofluorene (2-NFLU), 3-nitrofluoranthene (3-NFLUANTH), 9-nitroanthracene (9-NANTH), 1,5-dinitronaphthalene (1,5-DNNAPHT)—, 3 oxy-PAHs-2-fluorenecarboxaldehyde (2-FLUCHO), and 5,12-naphthacenequinone (5,12-NAPHTONA), it showed that the LC/MS arrangement provides a high degree of sensitivity and selectivity for the determination of these substances. In fact, it was demonstrated that it allowed the feasibility of its application to real samples. However, it was only possible to reliably report environmental levels of four of eight of these substances, at atmospheric concentrations between 0.01 and 240.62 ng/m^3, equivalent to 0.3 and 30 mg/g, respectively [56] (**Table 3**).

2.3.3 Two-dimensional gas chromatography (GC × GC)

The widespread use of capillary columns in the 1980s improved significantly the separation power of complex mixtures. This positioned gas chromatography as the technique of choice whenever analyzing volatile substances. However, it soon became clear that in some fields, the separation capacity offered by a single chromatographic column was not sufficient. For example, in the case of the oil industry, environmental applications or aroma analysis, whose separations are highly

Location	Compounds	Equipment	Ionization mode	References
Industrial area of Taranto, Italy	Nitro-PAHs in airborne, PM_{10}	GC/MS triple quadrupole	EI+	[48]
Seoul, Korea	PAHs in $PM_{2.5}$ airborne particles	GC-GC-TOFMS	ESI+, APPI+	[55]
Buenos Aires, Argentina	Oxy-PAHs and nitro-PAHs airborne, $PM_{2.5}$ and PM_{10}	UHPL-MS/MS triple quadrupole	APCI+	[53]
Zaragoza, Spain	PAH associated to the airborne particulate matter, PM_{10}	GC/MS triple quadrupole	EI+	[51]

Table 3.
Application of the GC/MS coupling for the analysis of PAH and its derivatives.

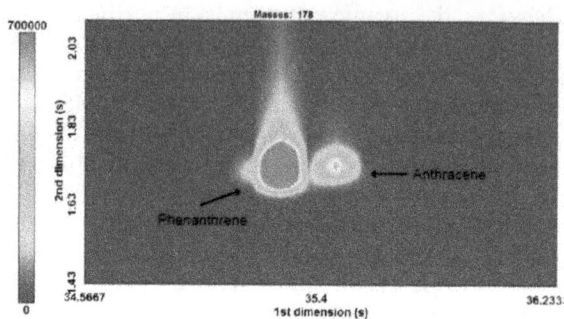

Figure 8.
Example of the application of the GC × GC for the separation and identification of structural isomers [61].

complex, often result in a chromatogram with a large portion of unresolved components [57]. Mass spectrometry can be used to resolve some of the complexity, but large concentration differences and structural isomers can complicate the spectral interpretation and data analysis. Some chromatographic resolution can be improved with an efficient, long, narrow bore, thin-film capillary column, but increased analysis time and decreased sample loading. This situation has been improved with the implementation of multidimensional gas chromatography (GC × GC). Multidimensional gas chromatography increases resolution by using two separate columns with two different stationary phases. One form of GC × GC is heart-cutting. After a preliminary evaluation of the sample, a portion of the unresolved GC effluent is reoriented to a different column before detection. Heart-cutting is a simple way to obtain a better separation of a complex mixture but just a portion of the one-dimensional separation can be improved with the second-dimension column. A high-frequency modulator is utilized by the comprehensive two-dimensional gas chromatography (GC × GC) for diverting the whole one-dimensional effluent onto a second-dimension column [58, 59] (**Figure 7**).

For instance, phenanthrene and anthracene are two important PAHs that can be used in order to assess whether material is petrogenic or pyrogenic in origin. Selecting mass 178 which is the molecular mass of the PAHs in question using the software allows isolated assessment (**Figure 8**).

As seen in the GC × GC, saving time in sample preparation, instrumental analysis, has the ability to analyze an extensive range of complex samples with the simultaneous target and nontarget detection, which makes it a powerful technique for the elucidation of complex matrices; however, it is expensive.

3. Conclusions

PAHs are one of the families of organic compounds associated with the airborne particles that have generated the most concern. Currently, there is evidence of the multiple impacts that these compounds have on human health and the environment. The exposure time to which humans are exposed, the concentration levels of PAHs in the air, as well as the phase in which they occur; that is to say gas or particle, and the size of the particles with which they are associated. All these parameters must be measured and determined by the appropriate methods of sampling, extraction, and analysis. In the last 20 years, analytical methodologies and equipment development have experienced significant advances; all this has allowed the advance of more selective and less destructive extraction procedures; in the same way, the purification methods of complex samples have been improved, but perhaps where greater progress has been made, has been in the instrumental analysis by coupling online extraction procedures, use of different detectors and the implementation of specialized software.

The main feature in the evolution of sampling systems has been the reduction of artifacts, through the use of adequate adsorption materials, as well as the mechanisms of PAHs uptake for both the gaseous and particulate phases. Regarding the extraction process, the greatest progress has been made in the reduction in the amount of solvents, compared with the traditional system (Soxleth). To reduce the multiple interferences of the extracts obtained from the airborne particles and to increase the sensitivity in the detection of the PAHs, several purification schemes have been implemented based mainly on the use of solid phase extraction cartridges. Finally, the chromatographic techniques are those that have experienced the greatest advances, starting with the GC–MS coupling. However, this configuration does not allow to distinguish between compounds that have structural isomers. What caused the coupling of several quadrupoles (QqQ) in the same equipment, thus increasing the resolution. Finally, the inclusion of two-dimensional chromatography GC × GC has allowed the simultaneous identification of compounds of different polarities, placing it as a powerful technique in the characterization of complex samples such as environmental samples.

There is no doubt the advance in the technology used for the analysis of PAHs in airborne particle matter in recent years. However, these new technologies require a high initial investment, in addition to highly qualified personnel, for this reason, before making the decision to acquire any of these new technologies, many aspects must be analyzed, for example if the equipment is going to be used for routine analysis, if the analytes to be studied can be extracted and detected with a simpler system, etc.

Acknowledgements

The authors wish to thank PRODEP (Program for the Development of Teachers), for the support in the financing of this publication.

Conflict of interest

The authors declare no conflict of interest.

Author details

Hugo Saldarriaga-Noreña[1]*, Rebecca López-Márquez[1],
Mario Alfonso Murillo-Tovar[2], Mónica Ivonne Arias-Montoya[1],
Jorge Antonio Guerrero-Álvarez[1] and Josefina Vergara-Sánchez[3]

1 Centro de Investigaciones Químicas, Instituto de Investigación en Ciencias Básicas
y Aplicadas, Universidad Autónoma del Estado de Morelos, Cuernavaca, Morelos,
Mexico

2 CONACYT-Centro de Investigaciones Químicas, Universidad Autónoma del
Estado de Morelos, Cuernavaca, Morelos, Mexico

3 Laboratorio de Análisis Ambiental y Sustentabilidad, Escuela Superior de Estudios
de Xalostoc, Universidad Autónoma del Estado de Morelos, Ayala, Morelos, Mexico

*Address all correspondence to: hsaldarriaga@uaem.mx

IntechOpen

References

[1] Wilhelm M, Ghosh JK, Su J, Cockburn M, Jerrett M, Ritz B. Trafficrelated air toxics and preterm birth: A population-based case-control study in Los Angeles County. California. Environmental Health. 2011;**10**:89-101

[2] Mohankumar S, Senthilkumar P. Particulate matter formation and its control methodologies for diesel engine: A comprehensive review. Renewable and Sustainable Energy Reviews. 2017; **80**:1227-1238

[3] Lamichhane S, Bal Krishna KC, Sarukkalige R. Polycyclic aromatic hydrocarbons (PAHs) removal by sorption: A review. Chemosphere. 2016; **148**:336-353

[4] Kim K-H, Jahan SA, Kabir E, Brown RJC. A review of airborne polycyclic aromatic hydrocarbons (PAHs) and their human health effects. Environment International. 2013;**60**:71-80

[5] Straif K, Baan R, Grosse Y, Secretan B, El Ghissassi F, Cogliano V. WHO International Agency for Research on Cancer Monograph Working Group. Carcinogenicity of polycyclic aromatic hydrocarbons. Lancet Oncology. 2005; **6**(12):931-932

[6] Ravindra K, Sokhi R, Van Grieken R, et al. Atmospheric polycyclic aromatic hydrocarbons: Source attribution, emission factors and regulation. Atmospheric Environment. 2008;**42**: 2895-2921

[7] Yamasaki H, Kuwata K, Miyamoto H. Effects of ambient temperature on aspects of airborne polycyclic aromatic hydrocarbons. Environmental Science & Technology. 1982;**16**:189-194

[8] Keller CD, Bidleman TF. Collection of airborne polycyclic aromatic hydrocarbons and other organics with a glass fiber filter-polyurethane foam system. Atmospheric Environment. 1984;**18**(4):837-845

[9] Araki Y, Tang N, Ohno M, Kameda T, Toriba A, Hayakawa K. Analysis of atmospheric polycyclic aromatic hydrocarbons and nitropolycyclic aromatic hydrocarbons in gas/particle phases separately collected by a high-volume air sampler equipped with a column packed with XAD-4 resin. Journal of Health Science. 2009; **55**(1):77-85

[10] Omar NYMJ, Abas MR, Ketuly KA, Tahir NM. Concentration of PAHs in atmospheric particles (PM10) and roadside soil particles collected in Kuala Lumpur, Malaysia. Atmospheric Environment. 2002;**36**:247-254

[11] Tang N, Hattori T, Taga R, Igarashi K, Yang XY, Tamura K, Kakimoto H, Mishukov VF, Toriba A, Kizu R, Hayakawa K. Polycyclic aromatic hydrocarbons and nitropolycyclic aromatic hydrocarbons in urban air particulates and their relationship to emission sources in the Pan-Japan Sea countries. Atmospheric Environment. 2005;**39**:5817-5826

[12] Tang L, Tang X, Zhu Y, Zheng M, Miao Q. Contamination of polycyclic aromatic hydrocarbons (PAHs) in urban soils in Beijing, China. Environment International. 2005;**31**:822

[13] Petty JD, Huckins JN, Zajicek JL. Application of semipermeable-membrane devices (SPMDs) as passive air samplers. Chemosphere. 1993;**27**: 1609-1624

[14] Wania F, Haugen JE, Lei YD, Mackay D. Temperature dependence of atmospheric concentrations of semivolatile organic compounds.

Environmental Science & Technology. 1998;**32**:1013-1021

[15] Harner T, Ky S, Genualdi S, Karpowicz J, Ahrens L, Mihele C, Schuster J, Charland J-P, Narayan J. Calibration and application of PUF disk passive air samplers for tracking polycyclic aromatic compounds (PACs). Atmospheric Environment. 2013;**75**: 123-128

[16] Hayakawa. Chromatographic methods for carcinogenic/mutagenic nitropolycyclic aromatic hydrocarbons. Biomedical Chromatography. 2000;**14**: 397-405

[17] Larsen RK, Baker JE. Source apportionment of polycyclic aromatic hydrocarbons in the urban atmosphere: A comparison of three methods. Environmental Science & Technology. 2003;**37**(9):1873-1881

[18] Dean JR, Xiong G. Extraction of organic pollutants from environmental matrices: Selection of extraction technique. Trends in Analytical Chemistry. 2000;**19**(9):553-564

[19] Havenga WJ, Rohwer ER. The use of SPME and GC-MS for the chemical characterisation and assessment of PAH pollution in aqueous environmental samples. International Journal of Environmental Analytical Chemistry. 2000;**78**(3–4):205-221

[20] Bodzek D, Luks-Betlej K, Warzecha L. Determination of particle associate polycyclic aromatic hydrocarbons in ambient air samples from the Upper Silesia region of Poland. Atmospheric Environment. 1993;**27A**:759-764

[21] Furton KG, Jolly E, Pentzke G. Recent advances in the analysis of polycyclic aromatic hydrocarbons and fullerenes. Journal of Chromatography A. 1993;**642**(1–2):33-45

[22] Escrivá C, Viana E, Moltó JC, Picó Y, Mañes J. Comparison of four methods for the determination of polycyclic aromatic hydrocarbons in airborne particulates. Journal of Chromatography A. 1994; **676**(2):375-388

[23] Bjorseth A, Lunde G. Long-range transport of polycyclic aromatic hydrocarbons. Atmospheric Environment. 1979;**13**:45-53

[24] May WE, Wise SA. Liquid chromatographic determination of polycyclic aromatic hydrocarbons in air particulate extracts. Analytical Chemistry. 1984;**56**(2):225-232

[25] Malloy TA, Goldfarb TD, Surico MTJ. PCDDS, PCDFs, PCBs, chlorophenols (CPs) and chlorobenzenes (CBzs) in samples from various types of composting facilities in the United States. Chemosphere. 1993; **27**(1):325-334

[26] Grossi G, Lichtig J, Kraul P. PCDD/ F, PCB and PAH content of Brazilian compost. Chemosphere. 1998;**37**: 2153-2160

[27] Meyer S, Cartellieri S, Steinhart H. Simultaneous determination of PAHs, hetero PAHs (N, S, O), and their degradation products in creosote contaminated soils. Method development, validation, and application to hazardous waste sites. Analytical Chemistry. 1999;**71**:4023-4029

[28] Spongberg AL, Witter JD. Pharmaceutical compounds in the wastewater process stream in Northwest Ohio. Science of the Total Environment. 2008;**397**:148-157

[29] Wang P, Yen VO, Tsai HT, Cheng JC. Determination of polycyclic aromatic hydrocarbons in airborne particulate by high performance liquid chromatography. Journal of the Chinese Chemical Society. 1988;**35**(1):13-21

[30] Yang X-J, Dang Z, Zhang F-L, Lin Z-Y, Zou M-Y, Tao X-Q, Lu G-N. Determination of decabrominated diphenyl ether in soils by Soxhlet extraction and high performance liquid chromatography. The Scientific World Journal. 2013;**2013**:840376

[31] Korcz W, Struciński P, Góralczyk K, Hernik A, Łyczewska M, Czaja K, Matuszak M, Minorczyk M, Ludwicki JK. Development and validation of a method for determination of selected polybrominated diphenyl ether congeners in household dust. Roczniki Państwowego Zakładu Higieny. 2014; **65**(2):93-100

[32] Chenxi W, Spongberg AL, Witter JD. Determination of the persistence of pharmaceuticals in biosolids using liquid-chromatography tandem mass spectrometry. Chemosphere. 2008;**73**:511-518

[33] Okuda T, Yamashita N, Tanaka H, Matsukawa H, Tanabe K. Development of extraction method of pharmaceuticals and their occurrences found in Japanese wastewater treatment plants. Environment International. 2009;**35**: 815-820

[34] Yu Y, Huang Q, Cui J, Zhang K, Tang C, Peng X. Determination of pharmaceuticals, steroid hormones, and endocrine-disrupting personal care products in sewage sludge by ultra-high-performance liquid chromatography-tandem mass spectrometry. Analytical and Bioanalytical Chemistry. 2011;**399**:891-902

[35] Frenich AG, Ocaña RM, Vidal JL. Determination of polycyclic aromatic hydrocarbons in airborne particulate matter by gas chromatography-triple quadrupole tandem mass spectrometry. Journal of AOAC International. 2010; **93**(1):284-294

[36] Gang LV, Wang L, Liu S, Li S. Determination of Perfluorinated

compounds in packaging materials and textiles using pressurized liquid extraction with gas chromatography-mass spectrometry. Analytical Sciences. 2009;**25**(3):425-429

[37] Lundstedt S, Bavel B, Haglund P, Tysklind M, Öberg L. Pressurised liquid extraction of polycyclic aromatic hydrocarbons from contaminated soils. Journal of Chromatography. 2000;**883** (1–2):151-162

[38] Piñeiro-Iglesias M, López-Mahía P, Vázquez-Blanco E, Muniategui-Lorenzo S, Prada-Rodríguez D, Fernández-Fernández E. Microwave assisted extraction of polycyclic aromatic hydrocarbons from atmospheric particulate samples. Fresenius' Journal of Analytical Chemistry. 2000;**367**:9-34

[39] Sibiya P, Chimuka L, Cukrowska E, Tutu H. Development and application of microwave assisted extraction (MAE) for the extraction of five polycyclic aromatic hydrocarbons in sediment samples in Johannesburg area, South Africa. Environmental Monitoring and Assessment. 2012;**185**(7):5537-5550

[40] Meyer A, Kleiböhmer W. Supercritical fluid extraction of polycyclic aromatic hydrocarbons from a marine sediment and analyte collection via liquid-solid trapping. Journal of Chromatography. 1993; **657**(2):31

[41] Kanagasabapathy VM, Bell RW, Yang P, Allan L, Au L, Parmar J, Lusis MA, Chapman RE. Developments in the supercritical fluid extraction of PAHs from environmental matrix. Journal of Chromatographic Science. 1995;**33**(8):467-474

[42] Vaz JM. Screening direct analysis of PAHS in atmospheric particulate matter with SPME. Talanta. 2003;**60**:687-693

[43] Chen P, Kang S, Li C, Rupakheti M, Yan F, Li Q, Ji Z, Zhang Q, Luo W,

Sillanpää M. Characteristics and sources of polycyclic aromatic hydrocarbons in atmospheric aerosols in the Kathmandu Valley, Nepal. Science of the Total Environment. 2015;**538**:86-92

[44] Zhang J, Wang J, Hua KP. The qualitative and quantitative source apportionments of polycyclic aromatic hydrocarbons in size dependent road deposited sediment. The Science of the Total Environment. 2015;**505**:90-101

[45] Somogyi Á. Mass spectrometry instrumentation and techniques. In: Vékey K, Telekes A, Vértes Á, editors. Medical Applications of Mass Spectrometry. Amsterdam: Elsevier; 2007

[46] American Chemical Society. Analytical Chemistry News & Feature. Washington, DC: American Chemical Society; 1996

[47] https://webbook.nist.gov/chemistry [Accessed: 05 June 2018]

[48] Hernandez F, Portoles T, Pitarch E, Lopez FJ, Beltran J, Vazquez C. Potential of gas chromatography coupled to triple quadrupole mass spectrometry for quantification and confirmation of organohalogen xenoestrogen compounds in human breast tissues. Analytical Chemistry. 2005;**77**:7662-7672

[49] https://en.wikipedia.org/wiki/User_talk:MiaJ601 [Accessed: 05 June 2018]

[50] Tutino MD. An improved method to determine PM-bound nitro-PAHs in ambient. Chemosphere. 2016;**161**:463-469

[51] Stashenho E. In: Chromatographica S, editor. GC-MS: Más de un Analizador de Masas, ¿para qué?. Vol. 2. 2010

[52] Murillo-Tovar M, Amador-Muñoz O, Villalobos-Pietrini RM. Selective separation of n-alkanes and PAH in complex organic mixtures extracted from airborne PM2.5. Chromatographia. 2010;**72**:913-921

[53] Callén ML. Nature and sources of particle associated polycyclic aromatic hydrocarbons (PAH) in the atmospheric environment of an urban area. Environmental Pollution. 2013;**183**:166-174

[54] Grosse SL. Liquid chromatography/atmospheric pressure ionization mass spectrometry with post column liquid mixing for the efficient determination of partially oxidized polycyclic aromatic hydrocarbons. Journal of Chromatography A. 2007;**1139**:75-83

[55] Lintelman JF. Determination of oxygenated polycyclic aromatic hydrocarbons in particulate matter using high-performance liquid chromatography-tandem mass spectrometry. Journal of Chromatography A. 2006;**1133**:241-247

[56] Fujiwara FG. UHPLC-(+)APCI-MS/MS determination of oxygenated and nitrated. Microchemical Journal. 2014;**116**:118-124

[57] Jung Hoon Choi JR-H-S. In-depth compositional analysis of water-soluble and -insoluble organic substances in fine (PM2.5) airborne particles using ultra-highresolution 15T FT-ICR MS and GC-GC TOFMS. Environmental Pollution. 2017;**225**:329-337

[58] Mondelo L, Lewis A, Bartle K. Multidimensional Chromatography. UK: John Wiley and Sons; 2002

[59] Liu Z, Phillips JB. Comprehensive two-dimensional gas chromatography using an on-column thermal modulator Interface. Journal of Chromatographic Science. 1991;**29**:227-231

[60] LECO, Simply GC × GC. 2017. https://www.leco.com/simply-gcxgc [Accessed: 05 June 2018]

[61] https://www.slideshare.net/KateJones7/kalin [Accessed: 05 June 2018]

Chapter 5

Petroleum Hydrocarbon and Living Organisms

Abdullah M. Alzahrani and Peramaiyan Rajendran

Abstract

Living matters are inadvertently exposed to the highly toxic petroleum hydrocarbon (PH) byproducts. Despite the fact that petroleum-related industry is globally thriving, the health hazard of most hydrocarbons is not well characterized. In human, organs and, sometimes, whole systems such as the nervous system, respiratory, circulatory, immune, reproductive, and endocrine systems are susceptible to PHs depending on the level of exposure. Marine organisms are known to be affected by PHs in various stages. Impacts from lethal to sub-lethal dose of PHs range from habitat destruction, mass mortality, and impaired physiological functions such as reduced feeding, slow growth and development, respiration problems, loss of locomotion, balance, and swimming ability. Bioaccumulation of toxic PHs in food chains in marine environment can be retained for decades and affect plants, animals, and eventually human. This chapter summarizes the PHs toxic effects on living organisms and the potential mechanisms of action based on epidemiological studies.

Keywords: hydrocarbon, petroleum, environment, toxicity, human, aquatic animals, aquatic plants

1. Introduction

Petroleum hydrocarbon (PH) is a term used to portray a wide group of a few hundred substance exacerbates that initially originate from raw crude oil. In this sense, PH is extremely a mixture of chemicals. They are called hydrocarbons in light of the fact that practically every one of them is made totally from hydrogen and carbon. Crude oils can move in the measure of each compound they contain the oil based products that are delivered utilizing crude oils. PHs are clear or light-hued fluids that vanish effectively and others are thick, dull fluids, or semi-solids that do not dissipate. Huge numbers of these items have trademark gasoline, kerosene, or oily scents. Since present day society utilizes such a significant number of oil-based items such as, gasoline, kerosene, fuel oil, mineral oil, and asphalt, pollution of nature by them is conceivably across the board. Pollution brought about by petroleum-based goods will contain an assortment of these hydrocarbons [1, 2].

Petroleum hydrocarbons spills are among the most broad and naturally harming contamination that are potentially dangerous to human and ecosystem health. Chronic exposure results from constant presentation to little proportions of oil over broad stretches of time [3, 4] and typically occurs in closeness to trademark spills, yet anthropogenic sources are also typical point sources, such as spilling pipelines, age discharges, or overflow from land-based sources can result in a solid inclination of high to low oil focus. Non-point sources, for example, atmospheric fallout and earthly spillover, additionally result in chronic exposure, yet may not contain an unmistakable inclination of focus. Hydrocarbons, as saturates, olefins, and aromatics, make up 97% of most petroleum [3]. Of these compounds, aromatics are among the most stable and may persist in the environment for long periods of time. Ceaseless exposures can result in subcellular impacts including altered metabolism, cell structure and function, or enhancement of chromosome mutation; this cascade of biological consequences associated with chronic pollution from frequent smaller spills are frequently viewed as a bigger risk than that related with acute exposure from tanker mishaps. Oil contamination in the ocean, regardless of whether from anthropogenic or common sources, endless or intense, is a noteworthy natural concern [2]. This chapter outlines the outdoor, occupational, and natural sources of PH exposure and considers the evidence relating to harmfully effect in living things.

2. Crude oil and its toxicity

2.1 Crude oil

Petroleum (crude oil) principally consist of carbon (83–87%) and hydrogen (12–14%) having complex hydrocarbon blend like paraffins, naphthenes, fragrant hydrocarbons, vaporous hydrocarbons (from CH_4 to C_4H_{10}). Other than these, crude oil likewise contains little measure of non-hydrocarbons (sulfur mixes, nitrogen mixes, and oxygen mixes) and minerals, heavier crudes contain higher sulfur [3]. Contingent upon power of hydrocarbons, petroleum is delegated paraffin base, middle of the road base or naphthenic base (**Table 1**). The unfavorable impacts of petroleum contamination on such necessary part of oceanic biological systems might be of extraordinary noteworthiness. Since the lethality of oil to biota is brought about by unsaturated hydrocarbons, naphthenic acids and another compound containing fragrant gatherings and nitrogen, the genuine harmful impacts portion is firmly identified with the measure of broke down non-unstable material [5]. The crude oil spills influence human well-being through their exposure to the intrinsic risky synthetics, for example, paraphenols and unpredictable benzene. The anticipated courses of introduction to synthetic compounds from the oil spill are inward breath, dermal contact, sustenance and water ingestion, and contact with the shoreline sand. This chronic exposure leads to affects physiological function such as hematologic, hepatic, respiratory, renal, and neurological functions.

2.2 Petroleum hydrocarbon

Oil energizes and oils are capricious mixes of hydrocarbons that move, among the fuel types, yet moreover inside each fuel type dependent upon maker, geographic zone, and customary use. The manifestations of these things are included a hardly any hundred hydrocarbon blends. Of these blends, toxicological information is available on only a not a lot of. This makes choosing the prosperity danger exposed by oil hydrocarbons troublesome.

Customarily, petroleum fuel or oil defiled destinations have been portrayed by two measures: explicit marker mixes called the synthetic compounds of concern (COCs) and by the aggregate of all the petroleum hydrocarbons called absolute petroleum hydrocarbons. The Petroleum Hydrocarbons Criteria Work Group (PHCWG) and the conditions of Washington and Massachusetts have created approaches that empower the improvement of human well-being hazard-based conclusion levels for PH. IDEM essentially concurs with these methodologies and has created comparable techniques. The PH conclusion levels depend on the

Hydrocarbons			
Hydrogen family	**Distinguishing characteristics**	**Major hydrocarbons**	**Explanations**
Paraffins (Alkanes)	Straight carbon chain	Methane, ethane, propane, butane, pentane, hexane	General formula C_nH_{2n+2}, boiling point increases as the number of carbon atom increases. With number of carbon 25–40, paraffin becomes waxy
Isoparaffins (Iso alkanes)	Branched carbon chain	Isobutane, isopentane, neopentane, isooctane	The number of possible isomers increases in geometric progression as the number of carbon atoms increases
Olefins (Alkenes)	One pair of carbon atoms	Ethylene, propylene	General formula C_nH_{2n}. Olefins are not present in crude oil, but are formed during process. Undesirable in the finished product because of their high reactivity. Low molecular weight olefins have good antiknock properties
Naphthenes	5 or 6 carbon atoms in ring	Cyclopentane, methyl cyclopentane, dimethyl cyclopentane, cyclohexane, 1,2 dimethyl cyclohexane	General formula $C_nH_{2n+2-2R_N}$. R_N is number of naphthenic ring The average crude oil contains about 50% by weight naphthenes. Naphthenes are modestly good components of gasoline
Aromatics	6 carbon atom in ring with three a round linkage	Benzene, toluene, xylene, ethyl benzene, cumene, naphthaline	Aromatics are not desirable in kerosene and lubricating oil. Benzene is carcinogenic and hence undesirable part of gasoline

Table 1.
Composition of petroleum.

non-malignancy end purposes of exposure. IDEM addresses the cancer-causing exposure by investigating certain cancer-causing COCs (benzene and certain cancer-causing polycyclic sweet-smelling hydrocarbons, cPAHs). Also, IDEM still requires source zone estimation of certain non-cancer-causing COCs (n-hexane, naphthalene, toluene, ethylbenzene, and xylene, in addition to non-cancer-causing polyaromatic hydrocarbons (PAHs) for waste oil). This new methodology separates the piece of explicit petroleum items into substance gatherings called portions, in view of carbon chain length and comparable physical/compound properties. Since the arrangement of each division is variable, and toxicological data is not accessible for each compound in each portion, the physical/concoction and toxicological properties of at least one surrogate mixes are chosen to represent each fraction [6].

2.3 Petroleum hydrocarbon effect on environment

Despite the fact that social and financial improvement generally relies upon petroleum hydrocarbon as it is an overwhelming wellspring of vitality, it has caused an enormous zone of defilement and significant unfriendly impacts. The defilement of petroleum hydrocarbon scatters from soil, water to human well-being. Petroleum hydrocarbon tainting of soil is a far reaching worldwide natural concern. Oil and fuel spills in soil are among the most broad and ecologically harming contamination issues as it is a threatening to human well-being and biological systems, particularly in cold

Product	Residential soils		Industrial soils	
	Direct contact (mg/kg)	Migration to ground water (mg/kg)	Direct contact (mg/kg)	Migration to ground water (mg/kg)
Gasoline range organics	3100	120	4300	1500
Diesel range organics	3100	230	5800	2300
High end hydrocarbon oils	3100	230	5800	2300

For more information, see the December 7, 2009, House Enrolled Act 1162 Interim Implementation Document at http://www.in.gov/idem/4202.htm.

Table 2.
PH closure levels.

Toxic effects	Plant species
Root development is reduced	Red beans (*Phaseolus nipponesis*) and corn (*Zea mays*)
A significant reduction in heights of seedlings, leaf length, and number of leaves	Soybean (*Glycine max*)
Significant reductions in plant height, leaf area and stem diameter was observed	Maize (*Zea mays* L.)
Hindered germination, reduced heights, and girths were observed	*Abelmoschus esculentus*
The plant growth was reduced significantly in low levels	Horsetail tree (*Casuarina equisetifolia*)
Crude oil pollution has an adverse effect on growth, yield, and leaf chlorophyll content	Air Potato (*Dioscorea bulbifera* L.)
Reduction in the length of the radicle for the four crop plants	*Arachis hypogaea*, *Vigna unguiculata*, *Sorghum bicolor*, and *Zea mays*

Table 3.
PH toxicity in plants.

area. Biochemical and physicochemical properties of soil is disintegrated by refinery items and it likewise restricts the development and improvement of plants. Water and oxygen shortfalls, just as to deficiency of accessible types of nitrogen and phosphorus, are the fundamental changes of soil properties because of tainting with petroleum-inferred substances [7]. Petroleum hydrocarbon sullied soil causing natural contamination of underground water which confines its use and causes financial misfortune, ecological issues, and diminishes the rural efficiency of the soil. Microorganisms, plants, creatures, and people are confronting helpless circumstance on account of the lethality of petroleum hydrocarbons. Soil compounds are one of the essential biotic segments which are in charge of soil biochemical responses. Petroleum hydrocarbon has unfriendly impacts of on soil enzymes activities (**Table 2**).

Oil spills influence plants by making conditions which make fundamental supplements like nitrogen and oxygen required for the plant development inaccessible to them [8]. Crude oil sullying at various dimensions caused critical decrease in the development of the plant utilizing plant tallness, crisp weight and leaf territory and the impact is relative to the dimensions of pollution [9]. Crude oil contamination has likewise unfriendly consequences for soil fruitfulness and plant generation. It could decrease or stop plant development prompting demise because of shaping a physical obstruction and covering the roots [10]. **Table 3** indicates antagonistic impacts of crude oil sullied soil in various plant species [7].

Petroleum hydrocarbon discharged in to the ocean, regularly amid transportation, prompting the contamination of a few destinations, and can in the long run achieve the coasts. Oil spills extending from low level releases to calamitous mishaps undermined beach front conditions; expansive spills generally are trailed by tidy up endeavors, yet total regulation is uncommon [7]. As dissolvability of petroleum hydrocarbon in water is commonly low, certain divisions of it drift in water and structure slim surface movies, which will encourage agglomeration of particles and regular natural issue, and effect on oxygen exchange. Other heavier portions will gather with the residue at the base of the water, which may influence base sustaining fish and living beings [7].

3. Petroleum hydrocarbon effect on living organism

3.1 Petroleum hydrocarbon on terrestrial animal

Animals are exposed to petroleum in many ways directly or indirectly. Some byproducts are formed during petroleum refining and processing which are used for the manufacturing of other products that are highly toxic. Constantly, these toxic compounds are inadvertently released into the environment and if this effect is connected to the effect of accidental crude oil spills worldwide, then these combined sources of unrestricted hydrocarbons constitute the major cause of environmental pollution. Despite the large number of hydrocarbons found in petroleum products, only a relatively small number of the compounds are well characterized for toxicity. Petroleum hydrocarbon molecules which have a wide distribution of molecular weights and boiling points cause diverse levels of toxicity to the environment **(Figure 1)**.

3.1.1 Uptake and metabolism

All together for oil metabolites have a direct organic impact on earthbound vertebrates, they should enter the individual, normally by means of ingestion, inward breath, or retention [11]. For most life forms, the essential course of PAH exposure in oil-influenced living spaces is through the ingestion of tainted soils, residue, and diet things. Thusly, species that feed vigorously on sediment-related invertebrates will generally be at more serious danger of PAH exposure in respect to higher order

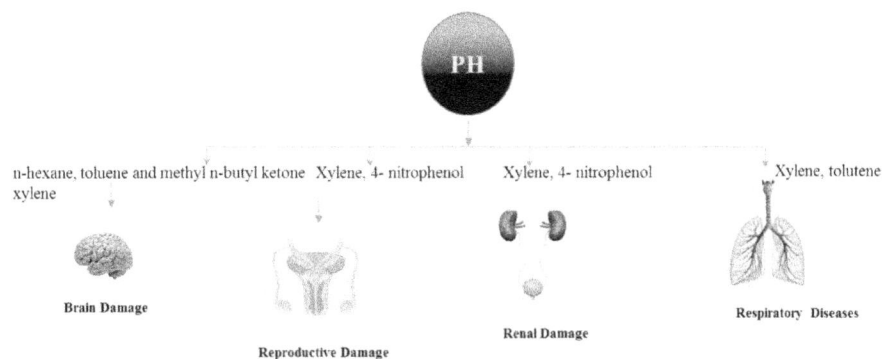

Figure 1.
PH effect on humans.

Figure 2.
Oil exposure in the environment. (a) Ear-tagged marsh rice rat, (b) tortoise, (c) seaside sparrow, (d) human, (e) oiled marsh (photographs: Philip C Stouffer). Adapted from [14].

consumers [12]. Be that as it may, PAHs only occasionally display sustenance web bioaccumulation and biomagnification; in this way, their potential for exchange up the natural pecking order is constrained. This is essentially connected with the expanded limit of vertebrates, including winged animals and warm-blooded creatures, to utilize and thusly dispense with PAH deposits.

PAHs can be perceived not long after presentation over a wide extent of vertebrate living creatures and tissues. For instance, field considers have recognized PAHs in the blood of feathered creatures and in turtle eggs and lab work has distinguished PAHs in snake skins [13]. Following their take-up, PAHs are processed by hepatic cytochrome P450 (CYP) oxygenase or blended capacity oxygenase proteins. Digestion can likewise happen in vivo. Because of this biotransformation, direct estimation of oil portions, for instance, hard and fast PAH in tissues isn't commonly a definite impression of exposure. Or maybe, the different isoforms of CYP (e.g., CYP1A) or CYP-related chemicals (e.g., ethoxyresorufin-O-deethylase [EROD]) that are upregulated within the sight of PAH are frequently utilized as roundabout biomarkers of crude oil or PAH exposure. For instance, hostage rodents presented to crude oil demonstrated a portion subordinate increment in a few hepatic CYP-connected chemicals. Field investigations of ocean ducks conceivably exposed to crude oil from the Exxon Valdez spill showed raised dimensions of these biomarkers in oiled regions even decades later (**Figure 2**) [14].

3.1.2 Toxicity in terrestrial vertebrates

Albeit molecular biomarkers, for example, CYP1A can be demonstrative of relative PAH exposure, only they may not suggest hurt or natural centrality. Unfriendly well-being impacts related with PAH exposure frequently result from the development of PAH metabolites, which have been exhibited to be genotoxic. In particular, these metabolites can tie to and harm DNA, framing DNA adducts (i.e., the official of DNA to a synthetic contaminant). For instance, hostage rodents presented to normally defiled soils with a wide scope of PAHs were found to have a subset of these PAHs in the liver and huge upregulation of EROD, and acceptance of DNA adducts came about. On the off chance that the DNA adduct is not fixed, generally typical cells can malfunction, prompting mutations and cancer. Other perceived lethal impacts of PAH on vertebrates incorporate conceptive broken-ness, immunosuppression, and edema. Be that as it may, a large portion of what is thought about PAH digestion originates from hostage ponders, in which dosing may not reflect characteristic dimensions. There are moderately few field investigations of harmfulness that connect physiological results with vertebrate exposure to PAHs. This is especially valid for earthbound species and the vast majority of this work has been led on feathered creatures. For example, an investigation of yellow-legged gulls (*Larus michahellis*) following the Prestige oil spill close Spain discovered blood parameters characteristic of hepatic and renal harm in grown-ups from oiled settle-ments, some of which associated with absolute PAH present in blood. Comparative work was led on marine species following the Exxon Valdez oil spill. Such impacts might be normal in earthly creatures of land and water, reptiles, and warm-blooded animals however remain inadequately considered [14–16].

3.2 Petroleum hydrocarbon on human

Synthetic concoctions and dispersants in crude oil can cause a wide scope of well-being impacts in individuals and natural life, contingent upon the dimension of presen-tation and helplessness. The Polyaromatic hydrocarbons are known parts of petroleum and petroleum-determined items. The PAHs are vital ecological toxins due to their cancer-causing nature. This mixes are routinely decided in modern waste water, drink-ing water, and groundwater. Guidelines on these lethal synthetic concoctions are as of now essentially in North America and Europe. The PAHs to a class of mixes with a high dangerous potential and thusly have a place with the gathering need contaminations. This dangerous synthetic substances can harm any organ system in the human body like the sensory system, respiratory system, circulatory system, immune system, regenera-tive system, tactile system, endocrine system, liver, kidney, and so on and subsequently can cause a wide scope of ailments and disarranges (**Figure 3**) [17, 18].

3.2.1 Petroleum hydrocarbon on nervous system

Long-term exposure to low levels of petroleum hydrocarbons may impair behav-iour and memory. This claim has led to an appraisal of the effect of these products on the nervous system.

3.2.1.1 PH on depression

A single exposure to a moderately high concentration of virtually any hydrocarbon solvent vapour will cause a general depression of CNS which, at high doses, will lead to unconsciousness. This property has been recognised for many years and some hydro-carbons (e.g. ethane) have been used as anaesthetics. Controlled short-term exposure of healthy subjects (up to 1-2 weeks), by repeated inhalation, to xylene, toluene, white spirit and jet fuel has shown that at levels of exposure above 250, 150, 300 and 200 ppm respectively an impairment of concentration, and of coordination occurs. These effects are readily and completely reversible on cessation of exposure [19].

3.2.1.2 PH on peripheral nervous damage

PH to delivering CNS melancholy n-hexane and methyl n-butyl ketone cause harm to the fringe nerves, especially of the feet and hands and this outcomes in

Figure 3.
Ways petroleum hydrocarbons (PHs) from oil enter aquatic animals.

unsettling influences of sensation and muscle shortcoming. In the event that the harm is serious, loss of motion may result; this loss of motion is infrequently lasting yet recuperation is moderate. Nerve harm of this sort has been found to happen in laborers with a background marked by genuinely overwhelming and delayed presentation to the dissolvable vapor and fluid. A common precedent is that of the shoemakers who worked for extended periods in restricted spaces and utilized a paste broke up in n-hexane. This kind of nerve harm has been appeared because of the arrangement of hexane 2,s-dione from the digestion of n-hexane and methyl n-butyl ketone and does not have all the earmarks of being brought about by different hydrocarbons or ketones; there is, in any case, proof that substances with structures identified with hexane or methyl n-butyl ketone can potentiate the nerve harm brought about by these two hydrocarbons [20].

3.2.1.3 PH on central nervous system

Over the most recent 10 years or something like that, various distributions have showed up, especially from Scandinavia which recommend that laborers utilized in occupations including presentation to natural solvents endure a disintegration in their enthusiastic parity, memory, knowledge, and forces of focus. They likewise portray a higher than normal occurrence of cerebral pains, wooziness, and other abstract grievances. Those examined were mainly painters and lacquerers. The condition has been given different names such as, painters' disorder, natural solvents illness, psycho-natural disorder, ceaseless Danish disorder, and incessant natural dissolvable inebriation. Specific examinations have given no proof of nerve or cerebrum harm in laborers influenced by this disorder. A basic assessment of these productions uncovered that the solvents primarily included are toluene, white soul, and fly fuel; xylene and styrene do not seem to have been involved. Toluene and white soul are imperative parts of numerous paints; what is more toluene is utilized widely as a deluding and cleaning specialist [21].

Examination utilizing a battery of mental tests in specialists who had been uncovered for quite a while to toluene vapor at air dimensions of around 100 mg L^{-1} neglected to build up any distinctions from controls. At high focuses, for example, it may happen in paste sniffing, toluene has been accounted to harm the cerebellum (a piece of the mind that controls balance). No investigations are, at present, accessible on the impacts of long haul introduction to white spirits, however two such examinations are accessible on stream fuel. Mental and mental tests uncovered no significant distinction between gatherings of laborers who had been presented to fly fuel for quite a long while and a coordinated control aggregate that had not been uncovered. Since painters and lacquerers are presented to an assortment of solvents, it is difficult to determine which of the solvents (or of the numerous blends accessible) is embroiled in the painters' syndrome.

All the more significantly, regardless of whether or to what degree introduction to solvents adds to the reason for this ailment is dubious. In most of the papers explored, lack of consideration has been given to the likelihood that different components could prompt the advancement of the discoveries in those uncovered. The most imperative of these variables are liquor addiction, utilization of psychoactive medications, introduction to lead or mercury, and propelling age. Lead and mercury are particularly relevant in this respect since they have, until recently, been important components of many paints; hence most of the painters who had been in this occupation for ten or more years must have had substantial exposure to these chemicals [22].

3.2.2 Petroleum hydrocarbon on respiratory system

Most of lung cancer patients examined had non-small cell lung cancer. A few investigations showed that a blend of volatile organic compounds (VOC) (benzene, xylene, toluene, and styrene), distinguished by GC-MS, could separate lung cancer patients from controls. By and large, the quantity of VOCs per demonstrate went from 7 to 33, with an affectability of 50–100% and a particularity of 80–100%. These examinations, together with studies exploring single VOCs, uncovered that the discriminative VOCs were prevalently alkanes (e.g., pentane, butane, and propane), alkane derivatives (e.g., propanol and various aldehydes), and benzene derivatives (e.g., ethyl-, propylbenzene). Albeit most VOCs levels were raised, certain dimensions (e.g., isoprene) were diminished in patients compared with controls. The indicative capability of VOCs profiles in lung cancer was additionally shown by gatherings the utilized eNose and other refined strategies. Additionally, breath profiles were diverse in patients with divergent histology (adenocarcinoma versus squamous cell carcinoma). Moreover, Peng et al. have shown unmistakable VOCs profiles in patients with lung, colon, bosom, and prostate cancer. The vital discoveries of VOC marks of various cancer types should be affirmed in more extensive clinical examinations. Multiple studies investigated the potential of VOCs to discriminate between lung cancer and other pulmonary diseases [23]. No single compounds (such as ethane), but a combination of multiple VOCs were able to distinguish lung cancer patients from patients with non-cancer pulmonary diseases (such as COPD, pleurisy, and idiopathic fibrosis) with a reasonable accuracy. Malignant pleural mesothelioma (MPM) is an uncommon tumor for the most part brought about by asbestos presentation. VOCs profiles had the capacity to analyze MPM in a group of subjects with long haul proficient asbestos presentation. In addition, Altomare et al. refined cyclohexane as conceivable marker of MPM [24].

3.2.2.1 Volatile organic compounds in other pulmonary diseases

Kanoh et al. exhibited that breathed out ethane was raised in patients with an interstitial lung malady (including sarcoidosis and idiopathic pulmonary fibrosis) compared with controls, with largest amounts in those with a functioning and dynamic infection. A small VOC, 2-pentylfuran, was ordinarily present in the breath of patients with a ceaseless pneumonic illness (including asthma and CF) with *Aspergillus fumigatus* in their respiratory examples, though this VOC was not recognized in the breath of controls [25]. Syhre et al. exhibited raised dimensions of breathed out methyl nicotinate in patients with pulmonary tuberculosis (TB) compared with solid controls. In a group of patients with doubt of TB, VOC designs had the capacity to recognize patients with TB from those without dynamic TB and healthy controls with a reasonable accuracy [26].

3.2.3 Petroleum hydrocarbon on reproductive system

Diesel fumes particulates (DEPs) have additionally been accounted to cause the disturbance of male conceptive capacity. Earlier examinations have demonstrated that DEP exposure aggravated spermatogenesis, bringing about decrease of every day sperm creation and motility, expanded morphological sperm variations from the norm, and ultrastructural changes in Leydig cells in mice. In male rodents, the guide-line of testicular capacity was adjusted bringing about height of serum testosterone and decrease of luteinizing hormone (LH) and sperm generation after DEP exposure. Scarcely, any epidemiological investigations have detailed regenerative harmfulness of xylene. In China, an examination was led on specialists who were presented to blended natural solvents in the petroleum business. The aftereffect of this examina-tion demonstrated that such a blend of natural solvents caused an expansion in the commonness of oligomenorrhea. There is another report on ladies uncovered solvents containing natural aliphatic and fragrant hydrocarbons. What's more, exposure to these solvents caused antagonistic result on regenerative hormones like decrease of pregnanediol 3-glucuronide (pd3G) in corpus luteum stage, pre-ovulatory lutein-izing hormone (LH), and estrone 3-glucuronide, and higher follicle stage pd3G. In addition, the commitment of xylene in the rate of such impact was more than 50%. Concerning barrenness, there are a few reports of abatement spermatozoa reasonabil-ity, and decline motility alongside lower acrosin activity discharge from spermatozoa which help in infiltration of the zona pellucida, diminished γ-glutamyl transferase activity, lactate dehydrogenase C4 (LDH-C4), and hoist the fructose level because of xylene exposure. One investigation showed that 4-nitrophenol (PNP) had estrogenic and antiandrogenic activities in vivo, prompting sterility. The amassing of PNP in air, water, and soil might be one factor in expanding frequency of sterility in people and creatures, yet epidemiologic examinations are pending [27].

3.2.4 Petroleum hydrocarbon on renal system

Different examinations have discovered an expanded sharpness of kidney tubules, declined creatinine in the pee and hematuria because of xylene exposure. Kidney impacts because of xylene were relied upon focus and portion which led to conglom-eration of m-xylene in the fats of kidney in the fringe. Other enzymatic exercises and expanded relative load of the kidney were additionally distinguished in rodents with various centralization of xylene. Histopathological assessment uncovered insignifi-cant ceaseless renal ailment. However, pee result was common, the essential unfa-vorable impacts identified was ascending in an adjustment in hyaline bead in male rodents and harm of kidney in the female rodents led to cell toxicity [28, 29].

4. Petroleum hydrocarbon on marine organisms

To begin with, oil spill mishaps could influence marine meteorological condition through scattering, disintegration, emulsification, and vanishing of the crude oil. When oil is spilled into the ocean, it could spread over the outside of the seawater. Some exploration revealed that a huge amount of spilled oil can frame 5×10^6 m^2 of smooth on the outside of the ocean water. The smooth could hinder the O_2/CO_2 trade straightforwardly and lead to an oxygen consumption and pH change in the ocean water. In this manner, a few reports demonstrated that marine desertification was brought about by oil spill mishaps. Moreover, oil spill mishaps seriously affect the marine/earthbound biological communities and human well-being. For instance, oil smooth structures an anaerobic condition in the ocean water and prompts the demise of widely varied vegetation. Oil spills can cause hypothermia of marine fowls and well evolved creatures by decreasing/decimating the protecting capacity of the plumage of feathered creatures and the hide of vertebrates. In the short-term, the poisonous establishes in petroleum could toxic substance or slaughter winged animals, well evolved creatures, angles and other marine living beings and harm the delicate sub-merged biological systems which lead to a horrible impact on the worldwide natural way of life, and in the long run mischief human well-being by harming inner organs, for example, kidneys, lungs, and liver. In addition, oil spill mishaps could influence marine plants and farming creation by blocking light and vaporous trade. It is evaluated that half of the all-out seaside wetland misfortune was brought about by oil spill mishaps. Finally, the oil contamination in marine condition can cause noteworthy financial misfortunes in the travel industry and marine asset businesses, for example, beach front salt industry, marine aquaculture, and fishery industry [30].

4.1 Fish

Oil spill mishaps could influence marine meteorological condition by means of scattering, disintegration, emulsification, and vanishing of the unrefined petroleum. When oil is spilled into the ocean, it could spread over the outside of the seawater. Some examination revealed that a huge amount of spilled oil can frame 59×10^6 m^2 of smooth on the surface of the ocean water [31, 32]. The smooth could obstruct the O_2/CO_2 trade straightforwardly and lead to an oxygen exhaustion and pH change in the ocean water. Furthermore, smooth could likewise impact water vanishing and precipitation in marine condition. Accordingly, a few reports showed that marine desertification was brought about by oil spill mishaps. Examines have exhibited expanded mortality of fish because of oil spills. Fish eggs and larvae are regularly powerless against poisonous oil mixes because of their little size, ineffectively created films and detoxification frameworks just as their situation in the water segment [33–35]. A research investigations have demonstrated that oil or oil mixes (for the most part polycyclic aromatic hydrocarbons, PAHs) at low fixations can execute or cause sub-deadly harm to angle eggs and larvae. Sub-deadly impacts incorporate, for example, morphological disfigurements, decreased sustaining, and development rates, and are probably going to build helplessness to predators and starvation. The few existing in situ investigations of fish mortality at spill locales demonstrate sub-deadly impacts or raised mortality of eggs and larvae [35].

4.2 Mammals

Marine warm-blooded creatures having an all-around created pelage would be relied upon to have oil stick promptly to them. This is upheld by research center examinations including ringed seals [36], ocean otters [37], and polar bears [38].

Extra proof incorporates the finding, in zones of spilled oil, of oil-fouled creatures, for example, harp seals (*Phoca groenlandica*) [39], grey seal (*Halichoerus grypus*) [40], and elephant seal puppies (*Mirounga angustirostris*) [41].

Exploratory introduction ponders in ringed seals and polar bears recognized that these species in any event had an incredible ability to discharge hydrocarbons collected from its exposure. The flood of ringed seals in an oil smooth brought about a take-up of hydrocarbons into tissues as examined beforehand, and likewise abnormal states in bile and pee. Renal and biliary discharge instruments gave off an impression of being successful to clear blood and most tissues of the gathered buildups by 7 days. Further, incredibly high buildup levels were found in pee following ingestion of a (14)C-naphthalene named oil [41]. A functioning digestion of the oil hydrocarbons, in any event of the aromatic parts, is demonstrated by the way that practically the majority of the (14)C-naphthalene movement was available as polar water-dissolvable buildups in both plasma and pee. Freedom of retained oil in polar bears appeared to happen by method for pee and bile [41]. Renal release officially huge in perspective on the high fixations and delayed nearness of oil hydrocarbons in pee, was presumably thought little of since the example readiness technique separated just dissolvable extractable hydrocarbons which were estimated by fluorometry, like the ringed seal oil inundation ponder. This strategy could not represent increasingly polar processed hydrocarbons. Albeit one may theorize that the biochemical instrument for hydrocarbon digestion in marine warm-blooded animals is like that of earthly vertebrates and depends on a blended capacity oxygenase framework, few subtleties of such a framework exist. Engelhardt [41] demonstrated that the chemical aryl hydrocarbon hydroxylase, one of the blended capacity oxygenases, exists in both liver and kidney tissues of ringed seals. Aryl hydrocarbon hydroxylase was observed to be inducible by in vivo exposure to unrefined petroleum, especially in kidney tissue where the movement of the compound multiplied.

5. Conclusion

Petroleum hydrocarbons may be high-profile events that can result in environmental impacts and affect the lives of living organisms. It is understandable that interest will be expressed by both individuals and organizations in knowing what damage was done and how long it will take to recover. However, while government agencies may have environmental quality monitoring programs in place for routine assessment, these will not be designed for large-scale pollution incidents. Government agencies that create guidelines for PH substances incorporate the EPA, the Nuclear Regulatory Commission (NRC), the Occupational Safety and Health Administration (OSHA), and the Food and Drug Administration (FDA). Proposals give profitable rules to ensure general well-being, however cannot be authorized by law. Government associations that create proposals for PH substances incorporate the Agency for Toxic Substances and Disease Registry (ATSDR), Centers for Disease Control and Prevention (CDC), and the National Institute for Occupational Safety and Health (NIOSH). Guidelines and proposals can be communicated in not-to-surpass levels in air, water, soil, or nourishment that are generally found on levels that influence creatures. At that point, they are acclimated to help ensure individuals. Now and then, these not-to-surpass levels contrast among government associations in view of various introduction times, the utilization of various creature contemplates, or different variables. Despite the fact that there are no bureaucratic guidelines or rules for PH by and large, the administration has created guidelines and rules for a portion of the PH parts and mixes. These are intended to shield people in general from the conceivable hurtful well-being impacts of these PH.

Petroleum Hydrocarbon and Living Organisms
DOI: http://dx.doi.org/10.5772/intechopen.86948

Author details

Abdullah M. Alzahrani and Peramaiyan Rajendran*
Biological Sciences Department, College of Science, King Faisal University, Hofouf,
Kingdom of Saudi Arabia

*Address all correspondence to: prajendran@kfu.edu.sa

IntechOpen

References

[1] Xu X, Liu W, Tian S, Wang W, Qi Q, Jiang P, et al. Petroleum hydrocarbon-degrading bacteria for the remediation of oil pollution under aerobic conditions. Frontiers in Microbiology. 2018;**9**:2885. DOI: 10.3389/fmicb.2018.02885

[2] Prince RC, McFarlin KM, Butler JD, Febbo EJ, Wang FC, Nedwed TJ. The primary biodegradation of dispersed crude oil in the sea. Chemosphere. 2013;**90**(2):521-526. DOI: 10.1016/j.chemosphere.2012.08.020

[3] Jiang W, Jinjun Z, Zhihui W, Yongxing Z. Correlations between emulsification behaviors of crude oil-water systems and crude oilcompositions. Journal of Petroleum Science and Engineering. 2016;**146**:1-9. DOI: 10.1016/j.petrol.2016.04.010

[4] Adenuga AA, Wright ME, Atkinson DB. Evaluation of the reactivity of exhaustfrom various biodiesel blends as a measure of possible oxidative effects: A concern for human exposure. Chemosphere. 2016;**147**:396-403. DOI: 10.1016/j.chemosphere.2015.12.074

[5] Muhammad A. Geochemical applications of polycyclic aromatic hydrocarbons in crude oils and sediments from Pakistan [thesis]. Lahore, Pakistan: Department of Chemistry, University of Engineering and Technology; 2010

[6] Gainer A, Bresee K, Hogan N, Siciliano SD. Advancing soil ecological risk assessments for petroleum hydrocarbon contaminated soils in Canada: Persistence, organic carbon normalization and relevance of species assemblages. Science of the Total Environment. 2019;**668**:400-410. DOI: 10.1016/j.scitotenv.2019.02.459

[7] Fowzia A, Fakhruddin ANM. A review on environmental contamination of petroleum hydrocarbons and its biodegradation. International Journal of Environmental Sciences & Natural Resources. 2018;**11**(3):555811. DOI: 10.19080/IJESNR.2018.11.555811

[8] Jyoti L, Smita C. Effect of diesel fuel contamination on seed germination and growth of four agricultural crops. Universal Journal of Environmental Research and Technology. 2012;**2**(4):311-317

[9] Tang J, Wang M, Wang F, Sun Q, Zhou Q. Eco-toxicity of petroleum hydrocarbon contaminated soil. Journal of Environmental Sciences. 2011;**23**(5):845-851

[10] Cermak JH, Stephenson GL, Birkholz D, Wang Z, Dixon DG. Toxicity of petroleum hydrocarbon distillates to soil organisms. Environmental Toxicology and Chemistry. 2010;**29**(12):2685-2694. DOI: 10.1002/etc.352

[11] Smith PN, Cobb G, Godard-Codding C, Hoff D, McMurry S, Rainwater T, et al. Contaminant exposure in terrestrial vertebrates. Environmental Pollution. 2007;**150**:41-64. DOI: 10.1016/j.envpol.2007.06.009

[12] Brooks AC, Gaskell PN, Maltby LL. Importance of prey and predator feeding behaviors for trophic transfer and secondary poisoning. Environmental Science & Technology. 2009;**43**: 7916-7923. DOI: 10.1021/es900747n

[13] Jones DE, Magnin-Bissel G, Holladay SD. Detection of polycyclic aromatic hydrocarbons in the shed skins of corn snakes (*Elaphe guttata*). Ecotoxicology and Environmental Safety. 2009;**72**:2033-2035. DOI: 10.1016/j.ecoenv.2008.11.001

[14] Christine MB, Jill AO, Stefan W, Philip CS, Taylor SS. Effects of oil on terrestrial vertebrates: Predicting

impacts of the Macondo blowout. Bioscience. 2014;**64**:820-828

[15] Oris JT, Roberts AP. Cytochrome P450 1A (CYP1A) as a Biomarker in Oil Spill Assessments. Cambridge, UK: Cambridge University Press; 2013. pp. 201-219

[16] Alonso-Alvarez C, Munilla I, López-Alonso M, Velando A. Sublethal toxicity of the prestige oil spill on yellow-legged gulls. Environment International. 2007;**33**:773-781

[17] Wu Q, Liu Z, Liang J, Kuo DTF, Chen S, Hu X, et al. Assessing pollution and risk of polycyclic aromatic hydrocarbons in sewage sludge from wastewater treatment plants in China's top coal-producing region. Environmental Monitoring and Assessment. 2019;**191**(2):102. DOI: 10.1007/s10661-019-7225-6

[18] Correa Pabón RE, Souza Filho CR, Oliveira WJ. Reflectance and imaging spectroscopy applied to detection of petroleum hydrocarbon pollution in bare soils. Science of the Total Environment. 2019;**1, 649**:1224-1236. DOI: 10.1016/j.scitotenv.2018.08.231

[19] Jones RD. Xylene/amitraz: A pharmacologic review and profile. Veterinary and Human Toxicology. 1990;**32**(5):446-448

[20] Wang C, Chen S, Wang Z. Electrophysiological follow-up of patients with chronic peripheral neuropathy induced by occupational intoxication with n-hexane. Cell Biochemistry and Biophysics. 2014;**70**(1):579-585. DOI: 10.1007/s12013-014-9959-7

[21] Riihimäki V, Savolainen K. Human exposure to m-xylene. Kinetics and acute effects on the central nervous system. The Annals of Occupational Hygiene. 1980;**23**(4):411-422

[22] Wang JD, Chen JD. Acute and chronic neurological symptoms among paint workers exposed to mixtures of organic solvents. Environmental Research. 1993;**61**(1):107-116

[23] Peng G, Hakim M, Broza YY, Billan S, Abdah-Bortnyak R, Kuten A, et al. Detection of lung, breast, colorectal, and prostate cancers from exhaled breath using a single array of nanosensors. British Journal of Cancer. 2010;**10, 103**(4):542-551. DOI: 10.1038/sj.bjc.6605810

[24] Altomare DF, Di Lena M, Porcelli F, Trizio L, Travaglio E, Tutino M, et al. Exhaled volatile organic compounds identify patients with colorectal cancer. The British Journal of Surgery. 2013;**100**(1):144-150. DOI: 10.1002/bjs.8942

[25] Kanoh S, Kobayashi H, Motoyoshi K. Exhaled ethane: An in vivo biomarker of lipid peroxidation in interstitial lung diseases. Chest. 2005;**128**(4):2387-2392. DOI: 10.1378/chest.128.4.2387

[26] Syhre M, Stephen TC. The scent of mycobacterium tuberculosis. Tuberculosis. 2008;**88**(4):317-323. DOI: 10.1016/j.tube.2008.01.002

[27] Krivoshto IN, Richards JR, Albertson TE, Derlet RW. The toxicity of diesel exhaust: Implications for primary care. Journal of American Board of Family Medicine. 2008;**21**(1):55-62. DOI: 10.3122/jabfm.2008.01.070139

[28] Kum C, Sekkin S, Kiral F, Akar F. Effects of xylene and formaldehyde inhalations on renal oxidative stress and some serum biochemical parameters in rats. Toxicology and Industrial Health. 2007;**23**(2):115-120. DOI: 10.1177/0748233707078218

[29] Monteiro M, Moreira N, Pinto J, Pires-Luís AS, Henrique R, Jerónimo C, et al. GC-MS metabolomics-based approach for the identification of a potential VOC-biomarker panel in the

urine of renal cell carcinoma patients. Journal of Cellular and Molecular Medicine. 2017;**21**(9):2092-2105. DOI: 10.1111/jcmm.13132

[30] Lee RF, Page DS. Petroleum hydrocarbons and their effects in subtidal regions after major oil spills. Marine Pollution Bulletin. 1997;**34**(11):928-940

[31] Gros J et al. First day of an oil spill on the open sea: Early mass transfers of hydrocarbons to air and water. Environmental Science & Technology. 2014;**48**(16):9400-9411

[32] Khan RA. Parasitism in marine fish after chronic exposure to petroleum hydrocarbons in the laboratory and to the Exxon Valdez oil spill. Bulletin of Environmental Contamination and Toxicology. 1990;**44**:59-763

[33] Fodrie FJ, Able KW, Galvez F, Heck KL, Jensen OP, Lopez-Duarte PC, et al. Integrating organismal and population responses of estuarine fishes in Macondo spill research. Bioscience. 2014;**64**:778-788. DOI: 10.1093/biosci/biu123

[34] Hjermann DØ, Melsom A, Dingsør GE, Durant JM, Eikeset AM, Roed LP, et al. Fish and oil in the Lofoten-Barents Sea system: Synoptic review of the effect of oil spills on fish populations. Marine Ecology Progress Series. 2007;**339**:283-299

[35] Langangen Ø, Olsen E, Stige LC, Ohlberger J, Yaragina NA, Vikebø FB, et al. The effects of oil spills on marine fish: Implications of spatial variation in natural mortality. Marine Pollution Bulletin. 2017;**119**(1):102-109. DOI: 10.1016/j.marpolbul.2017.03.037

[36] Geraci JR, Smith TG. Direct and indirect effects of oil on ringed seals (*Phoca hispida*) of the beaufort sea. Journal of Fisheries Research. 1976;**33**:1976-1984

[37] Siniff DB, Williams TD, Johnson AM, Garshelis DL. Experiments on the response of sea otters *Enhydra lutris* to oil contamination. Biological Conservation. 1982;**23**:261-272

[38] Oritsland NA, Engelhardt FR, Juck FA, Hurst RJ, Watts PD. Effect of Crude Oil Onpolar Bears. Environmental Studies No. 24. Dept. of Indian Affairs and Northern Development Canada; 1981. p. 268

[39] Warner RF. Environmental Effects of Oil Pollution in Canada: An Evaluation of Problems and Research Needs. Can. Wildl. Serv. Ms. Report No. 6451969. pp. 16-17

[40] Davis JE, Anderson SS. Effects of oil pollution on breeding grey seals. Marine Pollution Bulletin. 1976;**7**:115-118

[41] Engelhardt FR. Petroleum effects on marine mammals. Aquatic Toxicology. 1983;**4**(3):199-217

.